AF790671

Mobile Media and the
Change of Everyday Life

Joachim R. Höflich / Georg F. Kircher /
Christine Linke / Isabel Schlote (eds.)

Mobile Media and the Change of Everyday Life

PETER LANG
Frankfurt am Main · Berlin · Bern · Bruxelles · New York · Oxford · Wien

Bibliographic Information published by the Deutsche Nationalbibliothek
The Deutsche Nationalbibliothek lists this publication in the Deutsche Nationalbibliografie; detailed bibliographic data is available in the internet at http://dnb.d-nb.de.

Cover design and Illustration:
Olaf Glöckler, Atelier Platen, Friedberg

ISBN 978-3-631-59765-1

© Peter Lang GmbH
Internationaler Verlag der Wissenschaften
Frankfurt am Main 2010
All rights reserved.

All parts of this publication are protected by copyright. Any utilisation outside the strict limits of the copyright law, without the permission of the publisher, is forbidden and liable to prosecution. This applies in particular to reproductions, translations, microfilming, and storage and processing in electronic retrieval systems.

www.peterlang.de

Contents

Prospective uses and mobile media

Mobile communication and the change of everyday life: a short introduction

Joachim R. Höflich

A young couple with child has boarded the railway carriage. There are still seats left aboard the carriage. Immediately, a group of four seats with table is being occupied, subsequently; the luggage is being stowed away. Simultaneously, various utensils are being unpacked and placed on the table. Among the utensils are a mobile phone, a notebook as well as a children's book for reading. While the child, equipped with his bottle, walked the aisle up and down, and established contact with fellow-passengers, the young father flipped his notebook open and occupied himself with a shooter video-game. It was clearly recognizable, that he, equipped with his virtual weapon, swept his opponents out of the way. The player wore headphones, so that the game did not take up so much sound space. Meanwhile, the mother occupied herself with the mobile phone and may have sent a few text messages out. Then she briefly reached out for the book. But she was somehow bored – and she must have disliked, that her companion did not seem to want to share her some time. At least, she jolted every now and then the screen of the notebook in order to illustrate: 'Just stop it, at last!' Why this example? It shows exemplarily how public space is claimed for private settings – and that media are hereby an integral part of it. It virtually appears like an example of "living-roomisation" (Heine 2006) of public space and at the same time it is an example of how communicative everyday life patterns (here a communication in public space) have changed. Thus, the young couple demonstrated that they are living in a mediatized world and that media are used wherever one pleases (and, of course, wherever one can). That other people are co-involved herein does hardly seem to matter; third parties as onlookers and the handling of them belongs to the new behavioural repertoire of mobile media users. Whether this also affects domestic actions is another question (the young couple may quite oppose to all those who could witness these family happenings on the train and thus relate them to the couple's domestic situation).

It is generally assumed that new media change the established everyday practices. Euphoric people pin their hope on the fact that new media put people closer together. For sceptics it is quite the opposite – media isolate, estrange, and rob humans of their pivotal emphatic interpersonal competencies. And for both it is the media and more specifically technology that are in the deterministic sense of the word, causally responsible for this. However, what is actually the new thing, which is now changing the world so dramatically?

This fundamental question is also raised by Leopoldina Fortunati in her contribution "A discourse around theories on new media" in this volume: "Internet and the mobile phone are currently considered new media. But what are the media which are understood as new?" And she resumes: "These two examples show how the classification of new media continues to remain a very difficult problem. The definition of new media is not easier. (...) Here, for convenience, I will keep the mobile phone and the internet as examples of widely recognized new media. This choice however opens two questions: 1) the mobile phone and the internet are still to be considered new media, after more than one decade of their domestication and appropriation? 2) and if yes, to what extent new media are new?" Especially, when looking back when old media were still new (cf. Marvin 1988) a surge of euphoria and scepticism is displayed – and a phase of normalization proceeds, in which the insight arises, that a lot of which is regarded as particular new or even revolutionary, is basically the continuation of the previous with only new (media) means. The e-mail love-letter that turns out to be the revitalization of the bridal letters of the nineteenth century, the text message as the continuation of the telegram (Höflich 2003: 11). Though, with every media the established media practices change. Well, this may be known since Riepl and his, from all camps of communication science, appreciated 'law'. He writes that "as a ground rule of development in communication" follows that "the easiest means, forms and methods, when once established and found effective will neither be completely and permanently replaced nor put out of use not even by the most accomplished or most developed ones; but they rather preserve themselves alongside these, only that these means, forms and methods are forced to find themselves other applications" (Riepl 1913: 5). Looked at it in that light, an old media will always be re-invented – and not least this signifies, that every new media is new for every generation or respectively is discovered newly. Thus, it is not only the case that new media go alongside with a change in established practices, simultaneously; also the practices change when dealing with established media. And when one assumes that the significance of a media is defined through usage, thus the social significance of an 'old' media is time and again evaluated by this usage.

Since this volume dedicates itself to the subject of mobile communication and the transition in everyday life, one must not forget that mobile phones are embedded into a network of media practices respectively they are embedded into other media. The herein collected essays trace back to an international conference called "(Mobile) Media and the Change of Everyday Life" which was held at the German University of Erfurt in October 2008. The emphasis might have been placed on mobile communication, and thus an inventory of the research in this field should be targeted. However, since the term 'mobile' was put into parenthesis, the subject matter was meant to be considered in a holistic context. When picking out certain works hereinafter, which dedicate themselves to mobile communication, then, this does not only have pragmatic reasons[1]. When it comes to broach the issue of moments of transition, then two ways are mapped out: Either one specifies special transitions and traces them back to the technological or respectively media changes or one picks out a special innovation and pursues their consequences (cf. Lauer 1977: 159). The latter was pursued. However, when it comes to the subject of the context between mobile communication and everyday life, this time, this did not happen from the perspective of a media determinism according to which media are the principle cause of transition and set its direction (see also: Schroeder 2007). In order to overcome an unsophisticated determinism (may it be in the shape a technological, media-economic, social or cultural determinism), one has to presume a 'shaping of technology', with which it can be counted on choice and negotiability of technology (Williams / Edge 1996). As a roamer between the (scientific and creative-developing) camps Richard Harper, Professor and Principal Researcher at Microsoft Research Cambridge, shows with specific examples that (again and again) new ideas not necessarily are received like the developers (or should we say 'creators'?) have planned it. The idea of a 'shared editor' as a medium of organizational communication that enables to work on shared documents, was not accepted the way the inventors have planned it. The editor mutated into a game – they became devices for laughter and mischief because the playful character was in the forefront. "As I say, we found fun and utility in them through their playfulness, in supporting not professional engagement but our social selves – we were doing then pretty much what teenagers do now", says Richard Harper in his article "Absence to presence : a vision of the communicating human in computer-mediated communications technology research". Well, in one case, he describes, even a romance has evolved out of this. Nowadays, this would certainly not be so outlandish. We know this practice today as Instant Messaging!

In the meantime not only the question of the cause of the transition arises. When it already constitutes a problem to determine what is new with this new media one virtually faces a problem area when dealing with the question: what exactly is transition – and from which point on something is changing. "Maybe the most difficult question on the subject matter of social transition is the question of what transition actually is. This namely regards the often neglected question of what kind and what extent of change is to be regarded as social

1 However, it was not the first conference on mobile communication in Erfurt. Among others there should be mentioned especially conferences which were linked to publications: "Mobile Communication in Everyday Life" (Höflich / Hartmann 2006) as well as "After the Mobile Phone" (Hartmann / Rössler / Höflich 2008).

transition" (Randall / Strasser 1979: 23). Social change is the preferred field of sociology – however, the dispute is not necessarily a success story. It is even announced that the analysis of the social change is said to be one of the greatest weak spots in the theory formation of sociology (cf. i.e. Tjaden 1972) and not least because of ongoing critical dispute, it is annotated, and that it has disappeared from the agenda (cf. Müller / Schmid 1995: 25). It is surely indisputable that the social aspect of life is always subject to change. However, is every transition already a change? Generally, at least since Talcott Parsons it is distinguished between processes within a system and a transition of the system, whereas both are frequently (and for Parsons contributing to confusion) summed up as "dynamics". This distinction, however, does not change anything about the fact, that it is difficult to separate a change within the system from a change of the system or respectively to distinguish a switch from quantity into quality (Engels 1955: 53 et seq.). This does not change even when the consideration is done within the context of a concrete media as it is done here: the mobile communication. There is also the additional problem of locating the effect of single media within the context of comprehensive transition processes. Thus, it would be presumptuous to regard mobile phones as a causal force that changes the relation between private and public (a privatization of public space). Well, it contributes to that. However, this would not be possible, if other media, albeit not exclusively, had not yet made way for this transition of private and public. We herein focus especially on the change of the (communicative) everyday life. Everyday life is the unproblematic and taken for granted situation. It is "that realm of reality (…) that every attentive and normal adult with common sense would plainly assume as given" (Schütz / Luckmann 2003: 29; cf. also Gebhardt 2008: 100 et seq.). Change of everyday life would then be the evolving of new self-evidence. If this is the case then it is difficult to recognize since there is no problem cropping up. Thus, such studies are of importance that examine the transition right from the beginning and thus encounter eruptions which would only have overlapped in the process of banalification. Especially mobile communication, and especially the medium of mobile communication the mobile phone, has grown into a natural part of our lives. The studies presented here suggest (selected) fields, where transition has manifested itself starting from acting in public space to self-representation via fashion items.

These works reflect the difficulty of defining change, whereby different areas which changed because of the introduction of mobile phones can be detected. The first area is concerned with mobile phone use in public space. Transition shall hereby be regarded as "the transition of manners which define how actors interact with each other" (Schmid 1982: 105). Interactions are based on shared (expectable) acting situations, or to use Goffman's terminology: on shared frames of action, which indicate what is happening (cf. Goffman 1974). Each frame is in turn based on rules. Change would thus mean that frames change and with them the underlying normative basis for actions. This definition of change is certainly along the same line of Parsons' (1969: 43) definition who states that "the change of the structure of a social system means the change of its normative culture." Applied to media use, this means that previous frames and rules change and new ones, in the shape of developing media frames, are established. Such a frame-oriented perspective can be found in the contributions of Höflich and Kircher, as well as in Schlote and Linke in this volume. Both works are based on a research project on the topic of "Mobile communication, telematization of everyday life and the transition of media practices" which has been financed by the German Research Foundation (German: DFG Deutsche Forschungsgemeinschaft) up until June 2009 and which summarize important results of said study. The avid reader will hereby notice, that both contributions were allowed a little bit more space than usual. This was in no case unintentional, as the Erfurt conference was at the same time a forum, in which the results of these studies were presented to a (international) professional audience and put forward to discussion. We hereby want to thank the participants of the conference again for their constructive and critical remarks[2].

The public space is a social space and a socially regulated space, which again serves as social and normative context during mobile phone use. This is the basis of Joachim R. Höflich's and Georg F. Kircher's contribution "Moving and lingering: the mobile phone in public space" which is concerned with mobile phone use as well as moving and lingering in public space. Mobile phones are a nuisance factor. But people actually do not want to attract attention – neither in general nor when they use their mobile phones. It is not uncommon that people perceive the ringing of their mobile phones as embarrassing – or even stress-inducing (cf. Höflich 2009). This shows a "sense of place", but also a sense of privacy, which is not least reflected in distance behaviour. Thus, a feeling of discomfort is being caused by other people getting to close when one is just having a call on one's mobile phone. Furthermore, a gender-specific moment can be established. Women feel much earlier uncomfortable than men when other people get to close. Apart from gender, the type of relationship one has with each other is also important. At last, some places make one feel more uncomfortable than others. This again leads to the works of Isabel Schlote and Christine Linke: "Interaction and individual patterns of mobile communication in Public Places". The type of relationship is relevant, may it be in the case of our mobile correspondents or present third parties. In other words, mobile phone use is only half understood when one does not take into consideration the type of relationship. Another aspect which the authors bring to attention is the supra-situational moments of use, which again are reflected in concrete communi-

2 We would like to thank Britta Gutzeit und Berit Kuplent for the translation of the contributions of Joachim R. Höflich and Georg F. Kircher and of Isabel Schlote and Christine Linke. We also thank Bettina Schulz for her assistance with the print layout.

cation situations. This means, that mobile phone use is incorporated in people's activity patterns – "how people allocate their time to different activities and where activities take place" (Chapin 1974: 21). Relationships are already relevant when one is talking about (mutual) anticipation of activity patterns. Mobile phones hereby do not only help with the coordination of activities, but also with their encouragement, synchronization and adjustment. However, it is also shown, that such media interactions are incorporated into a complex web of considering present and non-present parties. Thus, the necessity arises to exceed a basic dyadic communication model and extend it into a model of mediatized mobile communication.

Satomi Sugiyama refers to fashion as a special area in her contribution "Fashion and the mobile phone: a study of symbolic meanings of mobile phone for college-age young people across cultures". Mobile phones are fashion accessories and vehicles for self-presentation. In the frame of an intercultural comparing study it was shown how important taste and style are, as well as the factor of time which goes along with the feeling of needing something new, "wanting new and improved features" and "wanting the newest phone". It further shows that not only the device but also its 'carrier' are subjected to evaluation. The transition of rules in relation to an appropriate appearance, an "appropriate appearance of the mobile phone at a particular point of time", seems to be going faster and faster. Thus strictly speaking, not the mobile phone is a cultural icon, but its appearance. There seems to be a trend towards a global generalization of the importance of mobile phones. But this represents only one side of the medal. The other one shows that subtle cultural differences – culture-specific differences – are emerging.

Jane Vincent writes in her article "Living with mobile phones" about British studies which have been conducted at the Digital World Research Centre since 2002. It presents the state of research with special regards towards the emotional moment during mobile phone use in connection with interpersonal relationships. Mobile phones are used not least to maintain social relationships. It thus serves as social and emotional glue. Dependences are shown when one is separated from one's mobile phone as thus the contact to people close to one's self seems to be cut which leads to a feeling of socio-emotional isolation. Mobile phones do not only symbolize the peers to which one feels attached to, but also deeper relationships. For example, the story of "Carl" is being told, who has lost a close friend and cannot bring himself to delete his phone number and the last text message he received from him. Thus, the friend's presence could somehow be ensured even two years after his death.

The relationship between emotions, technologically mediated communication and relationships is the topic of the essay "Mobile media and affectivity: some thoughts about the notion of affective bandwidth" by Amparo Lasén. The study of the term and the underlying definition of emotions shows how manifold emotions were understood and how isolated they are sometimes from any kind of rational action. The term emotional turn describes the transition from the supposition that emotions are individual and psychological features towards an understanding that they are fundamental aspects of social life. This understanding of emotions allows us to regard them not as private actions, but as an important help for orientation and decision-making in everyday life. From affective bandwidth approach we bridge to the aspect of media transportability of emotions. Hereby, different mobile media and their different applications are studied exemplary. At the same time, the common perception that technology-mediated communication has less affective bandwidth as person-to-person communication is being disagreed upon and it is made clear that these are merely different qualities of emotions, which cannot be categorized as being merely good or bad. Media enable new ways of emotional acting which can in certain situations and relationships be perceived as stress-reducing.

Furthermore, the mobile phone is perceived as a medium that overcomes boundaries. Thus, not only private and public, but also private and professional matters are merged. This is the topic of the contribution by Matthias Berg "Communicative mobility and mobile work: the management of everyday life and communication networks in a mediatized world". He is concerned with communicative practices in situations with job-related mobility against the backdrop of mediatization of everyday life. His work is based upon a concept of communicative mobility, with which a comprehensive mobility perspective is being presented which also exceeds use of mobile devices. A lot of case studies are presented which strongly insist on regarding media as a whole, meaning in a sense of comprehensively understood mobility as well as in a comprehensive media context with devices that serve "getting mobile" and stationary terminals with a "mobility focus".

The volume concludes with Rich Ling's and Pål Roe Sundsøy's article "The iPhone and mobile access to the internet". This contribution is making the collection round by pointing once more on the ongoing development of media technology and the ongoing change of mobile media practises in everyday life. Apple's iPhone was already a hit before it was even launched. It was however not only an insignia of status, but also a medium which was much more than just a telephone. This might hold true for mobile phones in general, but the iPhone stands for a new step towards mobile use of the internet. Thus, the authors ask whether or not the iPhone motivates people to go online via a mobile medium. This is, as has been found out, the case now. Going online has increased due to a usage convergence of the iPhone. However, this cannot be assigned to its technical features alone. There are moreover social and cultural factors that are taking effect here. It has shown that the age of users (which are predominantly male), as well as their professional context (with companies paying) are especially important. The study was conducted at a time when the iPhone was only available in the USA; thus it can be assumed that a pronounced internet affinity was transferred more easily to the new medium. At last, it could also be a self-fulfilling prophecy: Because of the fact that the iPhone was advertised as a mobile internet terminal, it was immediately understood as one by users.

All the articles collected in this volume show that mobile phone use and the transitions associated with it, can only be understood in concrete action contexts and that the transition of these contexts has to be analysed first. The examples show this with regards to public communication, people's activity patterns, communicative arrangements, fashion and self-presentation, relationships and emotional bonds, shift of borders from professional to private and last but by all means not least to certain technical platforms. Thus, it is again underlined that the mobile phone, like every other medium, shall not be reduced in a media deterministic sense. Only then it might be possible to show that it fits into a new transition that has already begun.

References

Chapin, F. Stuart (1974): Human Activity Patterns in the City. New York: Wiley & Sons.

Engels, Friedrich (1955): Dialektik der Natur. Berlin: Dietz Verlag.

Gebhardt, Julian (2008): Telekommunikatives Handeln im Alltag. Eine sozialphänomenologische Analyse interpersonaler Medienkommunikation. Wiesbaden: VS-Verlag.

Hartmann, Maren / Rössler, Patrick / Höflich, Joachim R. (eds.) (2008): After the Mobile Phone? Social Changes and the Development of Mobile Communication. Berlin: Frank & Timme.

Heine, Matthias (2006): Im öffentlichen Wohnzimmer. In: Frankfurter Allgemeine Sonntagszeitung, No. 26 (2006), 60-61.

Höflich, Joachim R. (2003): Einleitung: Mediatisierung des Alltags und der Wandel von Vermittlungskulturen. In: Höflich, Joachim R. / Gebhardt, Julian (eds.): Vermittlungskulturen im Wandel. Brief – E-mail – SMS. Frankfurt am Main: Peter Lang, 7-18.

Höflich, Joachim R. (2009): Mobile Phone Calls and Emotional Stress. In: Vincent, Jane / Fortunati, Leoplodina (eds.): Electronic Emotion. The Mediation of Emotion via Information and Communication Technologies. Oxford: Peter Lang, 63-83.

Höflich, Joachim R. / Hartmann, Maren (eds.) (2006): Mobile Communication in Everyday Life: Ethnographic Views, Observations and Reflections. Berlin: Frank & Timme.

Marvin, Carolyn (1988): When Old Technologies were New. Thinking about Electric Communication in the Late Nineteenth Century. Oxford: Oxford University Press.

Müller, Hans-Peter / Schmid, Michael (1995): Paradigm Lost? Von der Theorie sozialen Wandels zur Theorie dynamischer Systeme. In: Müller, Hans-Peter / Schmid, Michael (eds.): Sozialer Wandel. Modellbildung und theoretische Ansätze. Frankfurt am Main: Suhrkamp, 9-55.

Randall, Susan C. / Strasser, Hermann (1979): Zur Konzeptionalisierung des sozialen Wandels: Probleme der Definition, des empirischen Bezugs und der Erklärung. In: Strasser, Hermann / Randall, Susan C. (eds.): Einführung in die Theorien des sozialen Wandels. Darmstadt und Neuwied: Luchterhand, 23-50.

Lauer, Robert H. (1977): Perspectives on Social Change. 2nd ed. Boston: Allyn and Bacon.

Parsons, Talcott (1951): The Social System. New York: The Free Press.

Parsons, Talcott (1969): Das Problem des Strukturwandels: eine theoretische Skizze. In: Zapf, Wolfgang (ed.): Theorien sozialen Wandels. Köln, Berlin: Kiepenheuer & Witsch, 35-54.

Riepl, Wolfgang (1913): Das Nachrichtenwesen des Altertums. Mit besonderer Rücksicht auf die Römer. Berlin: Teubner.

Schütz, Alfred / Luckmann, Thomas (2003): Strukturen der Lebenswelt. Konstanz: UVK.

Schmid, Michael (1982): Theorie sozialen Wandels. Opladen: Westdeutscher Verlag.

Schroeder, Ralph (2007): Rethinking Science, Technology, and Social Change. Stanford, CA: Stanford University Press.

Tjaden, Karl Hermann (1972): Soziales System und sozialer Wandel. Stuttgart: Deutscher Taschenbuch Verlag.

Williams, Robin / Edge, David (1996): The Social Shaping of Technology. In: Research Policy, 25, 856-899.

Wyss, Eva L. (2003): Metamorphosen des Liebesbriefes im Internet. Eine korpusgestützte textlinguistische Bestimmung des Liebesbriefes und seiner Pendants im Internet. In: Höflich, Joachim R. / Gebhardt, Julian (eds.): Vermittlungskulturen im Wandel. Brief – E-Mail – SMS. Frankfurt am Main: Peter Lang, 199-231.

Theoretical discourses on mobile media

A discourse around theories on new media

Leopoldina Fortunati

New media: a difficult definition

The internet and the mobile phone are currently considered to be new media. But what are the media which are understood as new? Ithiel de Sola Pool (1990: 19) classified almost 25 devices as new media. Manovich (2001), when he tried to classify new media, cited websites, multimedia computers, electronic videogames, CD-ROM and DVD, and, at the end, virtual reality. Not a word about the mobile phone, for example, was mentioned. These two examples show how the classification of new media continues to remain a very difficult problem. The definition of new media is not easier. In fact, this notion is weak because, as Kim and Sawnhey (2002) point out, it is based on a concept of novelty, which is in itself a temporary concept. Here, for convenience, the mobile phone and the internet are kept as examples of widely recognized new media. This choice, however, opens two questions: 1) are the mobile phone and the internet still to be considered new media, after more than one decade of their domestication and appropriation, and 2) if yes, to what extent are new media new? The traditions from which the mobile phone and the internet come are quite different. Manovich (2001: 207) stresses that the mobile phone comes from the tradition of technologies of communication in "real time", while the computer / internet comes from the tradition of informatics and technologies of representation. But in the end, do new media represent an evolution of old media, or a true revolution? Van Dijk (1999) analyses this issue, recalling that Williams (1982) identified the second great concentration of innovation at the end of the 20th and beginning of the 21st century. According to him, we can speak of a revolution, but keeping in mind that a revolution, even if it represents a moment of discontinuity, a jump in respect to the past, also has elements of continuity with the past and elements of regression, not necessarily of positive progress and improvement (Fortunati, forthcoming). A further way to look at the definition of the mobile phone and the internet as new media, is to ask ourselves which among them is the newest technology, the technology which might represent the most important discontinuity in respect to the past. A time ago, it was proposed to consider not the internet, but the mobile phone as the most explosive technology (Fortunati / Contarello 2002; Fortunati 2007b). If new media is observed from a users' point of view, it is the mobile phone that expressed the highest speed in its diffusion, reaching, in less than 20 years, more than 3 billions people. This means that people have found this device the most convenient, friendly, and easy to use all around the world. Secondly, such a wide appropriation of the mobile phone means that people, by giving this preference to the mobile phone, have decided to reinforce body-to-body conversation through both voice and writing. Thirdly, the mobile phone has not only reduced the distance of technology from the human body, but has itself changed the direction of the development of technologies of information and communication, as is discussed below.

Van Dijk's suggestion, however, is a good indication, but is not sufficient to help us in properly answering these two questions. An attempt will be made to understand if focusing on the features and capabilities of new media could be a more effective strategy for answering these questions. An examination of the mobile phone and the internet has been chosen comparing their respective features: the portability / wearability, the structure of messages, the typology of language, the modality of use, and the infrastructure. While the mobile phone can be defined on the basis of one of its physical characteristics, such as its being a mobile or a portable and even a wearable technology (but which in time has happened to be used also in a sedentary way), the internet is more difficult to define from this point of view. The mobile phone is, in fact, a physical tool, while the internet traditionally needs a computer to exist and has to refer to the laptop for recuperating mobility. Both the mobile and the internet can be defined by the structure of the messages that are conveyed through them. The structure of messages in both cases can be circular, both synchronic (calls and skype) and a-synchronic (SMS and e-mail, instant messages, etc.), but also unidirectional (in the mobile phone: SMS news, horoscope, weather forecasting, and so on; while in the internet: online newspapers, websites, etc.) and even one-to-many direction messages (in the mobile phone the SMS that are sent to a list of friends, for example, on the occasion of a party, while for the internet there are mailing lists, bulletin board messages, blogs, forums, and so on). The mobile is, of course, stronger for voice calls and short mobile messages, while the internet for long, written information and communication. Another way to define the mobile phone and the internet is through the characteristics of the language they convey. The language can be oral, written, visual, audio (sounds or music), and mixed. Both the mobile phone and the internet are multimedia, even if the latter is more mature in respect to the multimedia language than the former. Furthermore, the definition of the mobile phone might be linked to the modality of its use, which is largely personal in industrialized countries, but which in many developing countries can be collective (Pertierra et al. 2002; Donner 2008). The use of the internet is less personalized. In the family, for example, usually more than one member uses the computer / internet and there are internet cafés in the public space.

New media can also be defined referring to the dimension of the infrastructure they require, which is that of the big media, according to Schramm's (1977) definition, since mobile phone and computer / internet systems require expensive, extensive and complex infrastructures. Another way to conceptualise new media could be through a comparison with old media, drawing upon their different, material characteristics. Ronald Rice et al. (1984: 35) in fact argue that new media, in contrast with old media, are those that allow or facilitate interactivity among e-actors or between e-actors and information. Van Dijk (1999) suggests that they are more interactive, more multimedia, more integrated, and more flexible in comparison to old media. But by producing such a comparison, we risk not understanding the very difference between old and new media, since from a certain point of view, points out Manovich (2001), new media could be seen as digitalized old media.

However, all these features (except perhaps for infrastructure) have become, in time, as such after an intense negotiation with the second, third and following waves of users (for the mobile phone see Fortunati 2007a). In fact, what are now the mobile phone and the internet owe a lot to the innovative role of e-actors who completely re-shaped them (Fortunati et al. 2010; Gebhardt et al. 2010). So, is there a way to look at the characteristics of the mobile phone and the internet as "new media", which takes into account not only their technological features, but also the human inside them? Is there a way to define new media focussing on e-actors and social, temporal, political, economic, and cultural contexts, as the tradition of users' studies would suggest, and contextualising the study of particular technological systems or media in a broader world of artefacts and culture, as the tradition of STS studies would suggest? From the first tradition, it would be necessary to assume what Raymond Williams (1975: 18 et seq.) has argued since 1975, that is that the evolution of technologies depends on actors' interests and that in this process, unpredictable or unintended consequences are not at all rare. In other words, he looked at social actors as an *independent variable* of the technological process, i.e. as the variable that might decide which technology will become a new one. From the second tradition, it would be sufficient to draw the main tenet, which in a certain sense follows from the first: if we consider social actors as an independent variable of the process, then it makes sense to assume a contextual perspective that underlines local practices and everyday life (Gerbner 1983; Bourdieu 1977), and that allows the discovery of unintended social consequences of media use.

Bockowski and Lievrouw (2008: 954), who try to bridge the studies of sociology of science and technologies with those of communication, face the same difficulties suffered when, with my colleagues Leslie Haddon, Gerard Claisse, Ulrich Lange, Josep Burcet and Anna Maria Manganelli, I tried to find a broader definition of the media which could include mass media and information and communication technologies (Fortunati 1998). At that time we opted for "telecommunications", while Bockowski and Lievrouw choose to speak of "media and information technologies". In Bockowski's and Lievrouw's definition, the list of devices to be included is still vague and the choice to eliminate "communication" from their definition is not so fortunate.

However, Bockowski's and Lievrouw's attempt to bridge two different fields of studies, such as those of communication and STS, is very timely, highly valuable and necessary. Their proposal to define "media and information technologies", as characterized by broad historical scope, infrastructural dimension, materiality, and interplay between this materiality and symbolic content and meaning, is sharable and, it seems to me, also extends to all the range of telecommunications. Moreover, they individuate three issues that have served as bridges between the two areas: causality in technology-society relationships (agency of either technological or societal factors), the technology development process (production-consumption), and the social consequences of technological change (evolution-revolution). These issues, they stress, have been framed, however, as binary oppositions between two opposite concepts, so they invite scholars to overcome the binaries in order to arrive at a more complex and deeper understanding. These issues are well framed in their work, but they are not exhaustive. For example, it strikes me that the categories time / space, which co-shape the social reality of media, are not indicated as a bridging issue. Another category I propose here as a driving principle, a law, is convenience. Convenience is one of the most important tenets supplied by new and old media, since it guides a lot of the social behaviour expressed in the domain of the use of new media. Under the apparent "rational" choice of the use of one device or another (i.e. economic or societal reasons), often the principle which is at work is that of the convenience, otherwise called Occam's razor.

It is important to fix some points of interest here: 1) it is necessary to bridge more fields: communication studies, mass media studies, STS, community informatics studies, social informatics studies, and ICT users' studies. Scholars cannot remain fragmented and divided in different areas, since they now work in a situation characterized by lack of information and knowledge, visibility of theories and issues, and results of research and interpretations; 2) on the basis of Bockowski's and Lievrouw's work, the main issues that cross all these fields and the theoretical tenets of the considered fields must be singled out and integrated: for example, the question of users' agency crosses all the fields; and 3) the novelty must be faced and inquiries made to define exactly what are new media.

New media as technological artefacts

A first step to achieve concreteness of the notion of new media taking into account e-actors' agency, is to revisit two notions that are central to the sociology of technology: the first is the notion of technological artefact, which is a key issue of the micro-sociology of technology, and the second is the notion of socio-technical systems, which is, instead, central to the macro-sociology of technology. Let us start by trying to conceptualise the mobile phone and the internet as technological artefacts. Firstly, it should be made clear that I propose to look at the mobile phone and the internet as technological artefacts, rather than as mediated communication, because considering mediated communication would have led to operating a reduction of the multiple factors implied in the social process developing around these media. The danger would have been not to acknowledge all the implications of the mobile phone and the internet as technological artefacts, and all the tasks and services which are conveyed through them. For example, in addition to conversation and information, the mobile phone also conveys orientation, localization, agenda, time measuring (clock) and alerting (alarm clock), mobile business, entertainment, m-learning, and so on, and the internet just keeps up.

While many studies have been produced on mediated communication (e.g. Whittaker, n. d.), an appropriate structural analysis of the specific characteristics of mobile communication compared to telephone communication and body-to-body communication is still lacking. Höflich is the one of those who did the main research in this direction (Höflich / Gebhardt 2005; Höflich / Hartmann 2006), but a substantial study capable of challenging the theories of mediated communication looking at the mobile phone is still to be done. Here, it would really be necessary to contrast mobile communication with telephone communication and body-to-body communication, but the main difficulty is represented by the discontinuity of the research on fixed and mobile communication carried out up to now. A robust attempt to investigate the ontological preconditions of mediated interpersonal communication from an e-actor point of view is that which has been carried out by Gebhardt (2010). By focusing on inter-subjectivity, understood as the problem of co-ordinating reciprocal communicative actions, this scholar referred to the socio-phenomenological concept of the "everyday life-world" as it was elaborated by Alfred Schütz (1967), and later on by Alfred Schütz and Thomas Luckmann (2003). He proposes to conceptualise communication technologies as communicative working tools, the interpersonal use of which is based on reciprocal orientations, rules and mutually established social practices.

Coming back to the proposal to see new media as technological artefacts, I draw on MacKenzie's and Wajcman's definition (1985) of technological artefacts. They understood these as physical objects, activities or processes, such as steel making, and know how going into their design. Clearly this is a definition of technological artefacts which remains inside the world of production of technologies. I propose to expand this notion by including the consumption world, which is also a laboratory of re-mediation and innovation of the technological artefacts. In this notion would be included, for example, e-actors' strategies of appropriation and domestication (Berker / Hartmann / Punie / Ward 2006), practices of innovation (Haddon et al. 2005), practices of use (Fortunati et al. 2010; Gebhardt et al. 2010), immaterial labour processes (Fortunati 2006, 2007c, 2009), societal and emotional structures (Vincent / Fortunati 2009), power relationships, and so on. All these dimensions are embodied in the transformations that e-actors impose on the technological artefact, and so reading an artefact could be an exercise which should allow the tracking down of e-actors' presence inside it. Now, if this definition of new media as technological artefacts is applied to the internet and the mobile phone, a concrete problem may arise. In fact, considering the physical body of these two technologies, it emerges that, while the mobile phone has a material and machinic body, the internet has not. The internet is like a saprophyte which needs the material body of other media, such as the computer, the mobile phone, or the television. It is this complex nature that has made it more difficult to analyse the internet as a technological artefact, while for the mobile phone an accurate analysis, even if only partial, already exists.

Some time ago, I analysed (Fortunati 2005a) a series of *design variants* on the material body of the mobile phones which were introduced by *e-actors*, tracing a map of them which is shortly summarized here:

a) the *individualization* of the mobile phone. The mobile phone is the first communication technology that has been put to a personal use and not a collective (familiar) one (Chiaro / Fortunati 1999);

b) its *sedentary use*. Despite the fact that the mobile phone was designed as a technology to be used while moving from one place to another, it has also been used at home, in the workplace, the restaurant, and so on;

c) the *overtaking of the landline phone*. Although it was designed as a complementary and secondary means of communication to the landline telephone, it rapidly overtook the fixed telephone;

d) its introduction of a *new dynamic between the private and public dimensions* by allowing people to shake up intimacy as a centripetal experience;

e) its *attraction of the private dimension costs from work communicative traffic*. Entrepreneurs, public administrations, and so on, have found a way, in many cases, of making workers use it for work without paying for the costs, while the boundaries between the sphere of work and social reproduction or civil society were being blurred (Fortunati 2006a);

f) *the use of text messages*. Conceived by the designers as a possibility for technicians to send each other short written messages during work, this function has been discovered and activated on a large scale by young people looking for a way of spending as little as possible to communicate (Ling 2003);

g) the use of *special ring signals* (or beeping). This new communicative register is linked to strong ritualism and a non-verbal language, which comes again from adolescents in industrialized countries (Fortunati / Manganelli 2003) and users in developing countries (Donner 2008), carried out for the purpose of communicating without paying;

h) the practice of deliberately *calling random numbers* in order to widen one's social circle, again introduced by adolescents;

i) the practice of *interacting with the media*. Voting during programmes by means of text messages, phoning while in the car, etc., have filled a previous gap in the relationship between audiences and the media and have resolved some technical problems for both media and users;

j) the use of the mobile phone *to present oneself*. The mobile phone is a fantastic resource for the construction of a public image, of self-assured and easy behaviour, to fill in gaps or empty moments with a precise rituality, to overcome timidity and discomfort in one's performance and to maintain one's image in a public space, with its anonymous crowds or threatening emptiness (Ling / Petersen 2005). Furthermore, women, when they are alone, in some situations use the strategy of demonstrating that they are using their mobile. As a form of self-defence, this display shows that they are connected to another person who can come to their rescue if need be;

k) the *aesthetic aspect* of the mobile phone. This tool is probably the communication technology that has most quickly transformed itself into a seductive and fashionable technological device; it stays so close to or on the human body that it has been the technology most destined to fall under the influence of fashion (Fortunati 1998b; Fortunati / Katz / Riccini 2003; Fortunati 2005b);

l) the attribution of *great importance to music* on the mobile phone. Starting with ring tones, but continuing with the possibility of listening to the radio on the mobile, young people in particular have imposed this additional form of convergence between music and the mobile phone.

The map of design variants coming from e-actors clearly speaks the language of society, of social and organizational structures, such as family and education system, of productive processes, such as those connected to immaterial labour processes, of the specific needs of relevant social groups, of social behaviour practices and ritualism, of common thinking, social learning and know how, and of the constellation of human relationships with objects, technologies included. All this goes, of course, to shape the body of the mobile phone (Pertierra 2007).

These user innovations and modifications of the mobile phone should be seen as the expression of consumer subjectivity, that is, of e-actors who have begun to see themselves as an independent variable of the whole economic process, trying to cut down costs and to adapt what they have bought to their needs and desires, thereby avoiding the command scheme embodied in every device, and applying all the modifications they consider suitable (the famous "dystonias of consumption"). Considering e-actors as an independent variable of the process of production of mediated communication, makes the effort carried out by researchers in assuming a contextual perspective capable of investigating local and everyday practices (Oudshoorn / Pinch 2003, 2008) intelligible and meaningful. The framework of the mobile phone as a technological artefact has opened the door to several theoretical perspectives. Studies in this vein also regarding the internet would be really welcome. The mobile phone has also been analysed in some studies as a dispositive item, in the revisitation that Agamben (2006) produced of this Foucaultian notion (Lasén 2010). Many studies, moreover, address the question of its being a cultural (Goggin 2006, 2008; Goggin / Hjorth 2009), emotional (Vincent 2003; Höflich 2009; Vincent / Fortunati 2009), fashionable (Fortunati 1998b; Fortunati / Katz / Riccini 2003; Fortunati 2005b) and symbolic object (Yung 2005). The mobile phone, just because it is an object of boundary on the merging of powerful dimensions, is able to reveal unexpected modalities of our current relationship with technological artefacts.

New media as socio-technical systems

A second and very promising step in conceptualising the mobile phone and the internet is through the lens of a major tenet of the macro-sociology of technology, which is the notion of a socio-technical system (Bijker / Hughes / Pinch 1987; Bijiker / Law 1992). Passing from a micro dimension (the artefact) to a macro dimension (the socio-technical system) is a good exercise, since, as Akrich (1992: 205) argues, artefacts "form part of a long chain of people, products, tools, machines, money and so forth" which needs to be investigated. The notion of a socio-technical system is a very useful framework in which to situate new media and investigate them in a whole scope of processes, practices and structures, characterized by contextualised space-time dimensions. The notion of socio-technical systems was tuned by the Tavistock Institute in London in the context of labour studies around the end of the 1950s. Ropohl (1979) argues that the concept of a socio-technical system was created to stress the interrelationship between humans and machines, as well as to understand the complexity of real situations, rather than at analysing separated aspects, or, I add, a single technological artefact. Although this construct was created to cope with the theoretical and practical problems of working conditions in industry, Ropohl suggests that the concept of a socio-technical system might be considered as the theoretical construct for reaching a comprehensive description and understanding of technology in its interrelationship with society. I argue that the socio-technical system is the ideal framework to study the complex interrelationship between society and new media, since people generally use a branch of technological artefacts in their everyday life. Furthermore, today it is even more useful to assume this concept, given the level of convergence supplied by media groups, telecoms, operators, content providers, etc., and the level of integration among technological artefacts that takes place in the consumption sphere of everyday life. This perspective is very similar to that expressed by Lievrouw and Livingstone (2006) in the introduction they wrote to the new edition of *The Handbook of New Media,* where they suggest seeing new media as made up of three components: the artefacts or devices used to communicate; the activities and the practices built around this use; and the social organization and the organizational forms which develop around these devices and their practices of use. They simply do not denominate new media as socio-technical systems, but their definition actually describes socio-technical systems. My personal vision, however, is to see the socio-technical system of new media as very flexible and highly dynamic, being based on processes and power relationships. To capture their main features, I would refer to a word launched by Bauman (2000) and which I find very appropriate: liquid. Liquidity and networking are the key words to describe the functioning of their organisation, instead of the classic term "hierarchic structure".

As sub-socio-technical systems, the mobile phone and the internet contribute, with all the other sub-socio-technical systems built around the other media, to the huge socio-technical system which forms the communicative environment in which we live, and that enhances, limits or complements body-to-body communication. Along with the other technologies, the mobile phone and the internet are the infrastructures of the communicative environment, and interact in the same everyday life setting. The relationship between e-actors and each technology (mobile phone and internet included) is embedded in a layer of networked relationships which build meanings, symbols, metaphors, dreams, attitudes, opinions, emotions, and behaviour (conflicts, resistance, appropriation, and innovation). This global socio-technical system, which includes all the technologies diffused in a given historical moment in workplaces in the public sphere as well in the domestic sphere, also interacts with other empires of signs and families of objects, and gives life to the material and immaterial structure of the socio-technical system. Analysing the mobile phone and the internet in that way becomes relevant, not only because e-actors are used to referring to a cluster of technologies in their everyday life, but also because of the increasing convergence among technologies and objects belonging to other families, such as textiles or biological elements, and even of the convergence between technology and social phenomena such as mobility, and so on (Fortunati 2008).

By the way, it is only by looking at new media as socio-technical systems that one of the most relevant novelties introduced by the mobile phone and the internet acquires the right relevance as a theme of reflection and debate: the reduction of the distance of the technological artefact from the human body. Other media, such as the radio and the walkman, had already reduced this distance from the human body, but these were not, in a certain sense, mainstream technologies. The computer / internet was the first to considerably reduce the distance of the human body from the screen, but the mobile phone has imposed a new rhythm in this escalation, arriving to hybridise itself with the human body (Fortunati / Katz / Riccini 2003). In this situation, the small optics (Virilio 1992), that is the distance between the subject and the object or other subjects, becomes the important point, since it is the distance which shapes the perception. Being the mobile phone on the human body, this device that it eliminates the distance between the subject who sees and the object which is seen, and puts in crisis not only the perception (which needs distance), but also the objectivity and, consequently, the reflexivity. Up to now, the technological process has always represented the extroflection of the communicative or work capacities of human beings from human beings themselves towards the environment surrounding them. The structural distance between devices and users was a great stimulus for social and personal reflexivity about technology. But now this distance has been radically reduced, especially by the mobile phone which has decidedly inverted this tendency.

Theories on new media

If the idea of bridging different fields of study – communication studies, mass media studies, STS, community informatics studies, social informatics studies, ICT users' studies – deserves to be developed, then a complex theoretical framework needs to be reconstructed. The theories on new media are actually middle range theories (Merton 1949) which represent an ensemble of hypotheses coming from the more influential schools of thought or research traditions, and have the ambition to look at the field of media with the purpose of capturing their main and general characteristics (Bourdon 2000). An interesting exercise of meta-theorisation of mass media is attributable to Umberto Eco (1964), and more recently to Bourdon, who in a certain sense develops Eco's model. Eco built two ideal-types, the apocalyptic and the integrated intellectuals, which were very useful to improve the intelligibility of the main features of studies on the mass culture. He distinguished between the intellectuals who saw especially the negative aspects of the mass culture and those who saw the positive aspects of the mass culture. Drawing on this, Bourdon developed a meta-model of the theories on mass media, distinguishing two couples of poles: the first couple opposes the scientific pole to the prophetic one, and the second, the optimistic (utopian) pole to the pessimistic (elite) pole. These two attempts are, however, tied in to a limited field of studies. In this case, it is still premature to try a meta-theorizing exercise on the theories connected to all the fields that should be taken into account. For the moment, let me try to produce a classification exercise of these theories, which I propose to classify in nine groups:

1) theories relative to new media and their features or capacities;
2) theories relative to their infrastructure;
3) theories relative to their diffusion / adoption;
4) theories relative to mediated communication;
5) theories relative to their content;
6) theories relative to users' agency, audiences and e-actors;
7) theories relative to innovation and design;
8) theories relative to the mutual shape between society and technology and
9) theories relative to the power and effects of media.

This classification is not exempt from the difficulty of identifying clear boundaries for which the different theories should be included in one group rather than in another. Secondly, the theories developed in all these domains are so numerous that it becomes very difficult to be exhaustive. Here, for reasons of space limitations, I will discuss only some significant examples of the first two groups.

Theories relative to new media and their features or capacities

What are the characteristics of new media? As was mentioned above, they are more interactive, more multimedia, and more integrated, so they have acquired to a greater measure the flexibility of human beings. To fix the features and capabilities of new media, I will discuss three significant contributions: Van Dijk (1999), Manovich (2001) and Lievrouw and Livingstone (2002). Van Dijk traces a useful framework to analyse the progressive interweaving among telecommunications, data information and mass communication, and the role played in this process by the spread of digitalization and broadband. He also examines the interactivity process in its bi-directional and multi-directional forms, in its degree of synchronicity, in the degree of control exercised by interlocutors, and of actions and reactions with a comprehension of the content and meaning. In his analysis, to describe the traffic models of information, Van Dijk applies the interesting model proposed by Bordewijk and Van Kaam in 1982 (cited in Van Dijk 1999): allocution, consultation, recording, and conversation. But he also argues that the raising of integrated networks implies a combination of these models in a unique medium. Finally, in order to answer the question "What can we do with new media?", he proposes a new theory on the capacity of communication of new media. This capacity is measured on the basis of nine elements: speed, geographical extension, social adoption and appropriation, capacity of memory, accuracy, selectivity, interactivity, richness of stimuli, complexity and protection of privacy.

Manovich (2001), however, offers a more powerful example of a theory about new media which has a heuristic value too. New media, he writes, represent the point of convergence of two different traditions: informatics and media technologies. Media technologies in their turn have developed along two different trajectories. The first is that of representation, which includes film, audio and video cassettes and various digital formats, while the second is that of communication in real time. The new cultural forms of the last century, such as the radio and the television, come from the merging of these two trajectories. The mobile phone comes instead from the communication in real time, which makes it no longer necessary to be physically present in a certain place to affect the surrounding reality (tele-presence as anti-presence), while the internet comes from informatics and media technologies.

Drawing again on Manovich, the archetypal machine of informatics is Babbage's analytic machine, and that of media technologies, is Daguerre's daguerreotype. Unfortunately, this turns out to be a mistake, since the daguerreotype does not allow production in series[1]. Manovich should have referred to the calotype, which is the true archetype of the photograph and of new media. Its procedure is based on the principle of negative-positive, and, differently from the daguerreotype, it allows the printing of a virtually unlimited number of prints (positive) from a unique negative. But apart from this mistake, if we operate the necessary correction to Manovich's analysis, his analysis remains fundamentally valid. These two machines, and here I would say Babbage's analytic machine and the calotype, were fundamental for the functioning of the modern mass societies as they allowed the possibility of spreading and strengthening the same ideology to millions of citizens, and to record and retrieve the fundamental data of society. Manovich's theory does not focus on the technical characteristics and capabilities of new media, which are visible on the surface of their body, as Van Dijk did, but on the normative principles that have inspired their creation and that govern them in an invisible way. He proposes five principles, which are: the *numeric representation*, the *modularity*, the *automation*, the *variability*, and the *cultural trans-codification*. The first and the more important principle is the numeric representation, that is the fact that new media are composed by a digital code, can be described in mathematical terms, are subject to algorithmic manipulation, and so are programmable. The programmability is the true watershed between the old and new media. The conversion of continuous data in a numeric representation is

1 This is pointed out by Marco Amato: http://www.codablu.eu/ (retrieved February 15, 2009).

called digitalization, which is made up of two stages: sampling and quantification. The sampling transforms continuous data into discontinuous data (discreet), relative to distinct units, and in a certain sense it embodies the logics of language which by definition is discontinuous in its structure. The frequency of sampling is called resolution. Almost all new media require the combination between continuous and discreet codification. The second principle, the modularity, which Manovich calls "the fractal structure of new media", implies that new media might be seen as organic ensembles of discreet samplings (pixel, voxels, characters, scripts). The third principle, the automation, that is the numeric codification of new media and their modular structure, allows the automation of many operations, which are necessary for the creation, manipulation and access to the media. This has the very important consequence that human intention can be removed, at least in part, from the process, with positive and negative consequences. Positive, since automation might eliminate human errors in the production process, negative since automation makes human involvement, responsibility and control in many stages of the production process superfluous. At the same time, automation allows the efficient retrieval, organization and access to all the material accumulated over the last 150 years. The fourth principle, the variability, means that new media are mutable and liquid. While old media were sons of the industrial revolution which had allowed the possibility of technically reproducing the work of art (Benjamin 1939), new media are, as was said before, sons of the post-industrial era for mass societies. So they allow not only the reproduction of a lot of identical copies of a medial object, but they also allow the reproduction of a lot of different versions of it. Secondly, they allow the fact that these different versions, instead of being integrally created by a human being, can be assembled by a computer itself. Manovich points out rightly that the logic of new media corresponds to the post-industrial logic of production on demand and just in time. The principle of variability is a useful framework which allows the study of several characteristics of new media until now studied in a separated way: namely, the existence of a database; the fact that it is possible to separate content from interface; the automatic personalization of the composition of the medial object; the ramified interactivity, and also the open interactivity; the hypermedia; the use of the periodical updates; and the scalability, i.e. the possibility to generate versions of the same medial object in various dimensions. It is evident that new media are, on the whole, generated by a society in which, at a certain point, people have refused the structural consequences of the massification process, such as homogenization, indifferentiation and standardization, and have contrasted these consequences by opting for a development of individualization. This has translated into a strong challenge of post-industrial society which has to be able to personalize its commodities, passing by mass scale to individual scale production. The variability guarantees users, in particular, that their choices in the practices of use of new media can be unique, instead of being pre-programmed and common to all the users. But, as Manovich rightly argues, a main effect of the principles of automation and variability is that users now have more power of control and determination of the representation of the world in the medial object, but they pay for this major freedom and power with their work. Users offer information, news, pictures, articles, and so on: they work without being paid. The example of the automatic switchboards reported by Manovich is very appropriate. In this case, the work is done by customers instead of by employees. If in previous times a customer could have access to information by talking to an employee, now he / she is obliged to invest time and energy surfing through a series of menus to obtain the same result. The fifth principle, the trans-codification, describes the fact that new media make it easier to pass from one format to another, at least at a technological level.

So, in Manovich's analysis, new media appear characterized by these five normative principles, among which the programmability is the most important one. These principles express a new way to create the illusion of reality, to manage the imagination, to address users, and to organize space and time. New media, sons of post-modern society, have developed thanks to a combination of new technological innovations and a new maturity of users and audiences. Better still, they are still insufficient in respect to e-actors' agency, since e-actors on average are more educated and have a decennial experience as an audience. On the basis of these normative principles, Manovich is able to propose a second order definition of new media: they are discreet, have in common the same digital code (the computer works as a multimedia lector), allow a random access and, being subject to digitalization, lose information, allow coping without any qualitative loss, and are interactive. A limit of this second order definition, however, is that the concept of digitalization is ambiguous, and the concept of interactivity remains vague.

The third more recent approach is that proposed by Lievrouw and Livingstone (2002) in their work *The Handbook of New Media.* Here the editors in the new introduction (2006) come back to the old definition of ICTs (technologies of communication and information), to which they add "and their relative social contexts". However, they try to understand in which way new media, their practices and their social organizations are different from those related to the previous media. The main result is that while the studies and research on the latter were characterized by a tripartite structure, including production, text and audience, the structure of the former is characterized by terms such as artefacts, practices and social organization. What is also important for these two authors is that while the tradition of mass communication has for a long time dealt with the linear relationship between production, text and audience, in the research on new media, the social shaping and the social consequences, they suggest, should be seen as an ensemble, and the uses should be analyzed along with interactivity, reconfiguration and reciprocity. Lievrouw and Livingstone (2002) refer to Bolter's and Grusin's (1999) work, according to which there is a continuous process of re-mediation in which old media are reshaped and absorbed by new media, both shaping reciprocally.

As to the social consequences of new media, Lievrouw and Livingstone (2002) indicate ubiquity, interactivity and the network structure as the most important. The ubiquity is the idea according to which new media influence everybody in the societies in which they are used, irrespective of the fact that not all people use them. Interactivity is the idea according to which users have more degrees of freedom and power in their relationship with new media. The network structure of new media makes them increasingly different from the hierarchical configurations and distribution logics of old media. Among the main social consequences of new media, however, Jenkins' (2006) work on convergence and all the debate on this topic collected by Kristof Nyiri (2008), and especially the fact, underlined again by Lievrouw and Livingstone (2002), that the current convergence among ICTs has been facilitated and shaped by the parallel convergence of entertainment, education, labour, civic activities and interpersonal communication should not be ignored.

Theories relative to the infrastructure of new media

There is a small circulation of studies on theories relative to the *infrastructure* of new media inside the field of communication studies and of mass media studies. They are more frequent in STS studies, but are especially focused on the internet infrastructure. Curiously, mobile phone infrastructure is studied and debated in other areas, such as organization, economic and information studies. A seminal study on the role of infrastructure in the implementation of new media is that carried out by Star and Bowker (2006). If it has become increasingly popular among scholars to understand new media not only as technical, but also as social, cultural and economic elements, the infrastructure, such as the mobile phone network, broadband, or wireless infrastructure, still has difficulty to be seen as something which is not merely technological. Star and Bowker point out instead that the infrastructure should be understood as a relational and social notion, since it stays in the middle between people and devices. David (1995) shows how infrastructures are rooted in many ways in organizations. The main characteristics of infrastructures are described as: deeply rooted in other structures, social organizations and technologies; transparent, in the sense that they are invisible and become visible only when they do not work; having a broad field of action, in the sense that they are generally something related to society instead of to communities; part of a community of practices, since strangers have to familiarize with them; being often connected to customs and their inertia, and are generally built on pre-existing elements or other infrastructures; and, finally, standardized. The socio-technical aspects of the infrastructure are also well illustrated by Hughes' work (1983), which shows how a technical problem may find a political solution. However, the possible political solution is not casual, since infrastructures are generally public equipment, so they are not a property either of industries or of citizens. Thus they are one of the fields of public bodies and administration, state and politics. They are a premise of the economics and logistics of a country, serving all the social actors. They can be local or national, but of course they are more national than local. When they exist at national level, they are "the great works" that states are supposed to guarantee to citizens.

After being installed, an infrastructure transforms itself sometimes rapidly, at other times in a discontinuous way (think of the countries which went directly to cellular networks without passing through fixed lines). It would, however, always be important to look also at this level, understood as a social issue. A systematic collection of the studies done on the internet and the mobile phone infrastructure could be very useful.

Conclusion

New media pose several theoretical challenges starting with those connected to their definition and classification, and continuing through the necessity to develop future studies both at micro- and macro-sociological level by exploring further the notion of new media as technological artefacts and as socio-technical systems. Another challenge is related to the necessity of merging many different fields of study, such as communication studies, mass media studies, STS, community informatics studies, social informatics studies, and ICT users' studies, and to properly manage the incredible theoretical richness that has been developed in these fields of study. The multidisciplinary approaches, which by vocation put very different fields in contact, are particularly relevant for studying the complexity of new media, but they often present theoretical and methodological problems. To understand new media, we have nine different groups of theories to be assessed and integrated in order to progress. It is time to do so also to be able to carry out more sophisticated empirical research.

References

Agamben, Giorgio (2006): Che cos'è un dispositivo. Roma: Nottetempo.

Akrich, Madeline (1992): The De-Scription of Technical Objects. In: Bijker, Wiebe E. / Law, John (eds.): Shaping Technology / Building Society. Studies in Sociotechnical Change. Cambridge, MA: MIT Press, 205-224.

Bauman, Zygmunt (2000): Liquid Modernity. Cambridge: Polity Press.
Benjamin, Walter (1939): The Work of Art in the Age of Mechanical Reproduction. In: Mast, Gerald / Cohen, Marshall (eds.) (1976): Film Theory and Criticism. London, New York, Toronto: Oxford University Press, 612-634.
Berker, Thomas / Hartmann, Maren / Punie, Yves / Ward, Katie J. (eds.) (2006): Domestication of Media and Technology. Maidenhead: Open University Press.
Bijker, Wiebe E. / Hughes, Thomas P. / Pinch, Trevor J. (1987): The Social Construction of Technical Systems. News Directions in the Sociology and History of Technology. Cambridge, MA: MIT Press.
Bijker, Wiebe E. / Law, John (1992): Shaping Technology / Building Society. Studies in Sociotechnical Change. Cambridge, MA: MIT Press.
Boczkowski, Pablo / Lievrouw, Leah A. (2008): Bridging STS and Communication Studies: Research on Media and Information Technologies. In: Hackett, Edward J. / Amsterdamska, Olga / Lynch, Michael / Wajcman, Judy (eds.): New Handbook of Science and Technologies Studies. Cambridge, MA: MIT Press, 949-977.
Bolter, Jay David / Grusin, Richard (1999): Remediation: Understanding New Media. Cambridge, MA: MIT Press.
Bourdieu, Pierre (1977): Outline of a Theory of Practice. Cambridge: Cambridge University Press.
Bourdon, Jérôme (2000): Introduction aux médias. Paris: Editions Montchrestien, E.J.A.
Chiaro, Marina / Fortunati, Leopoldina (1999): Nouvelles technologies et compétences des usagers. In: Réseaux, 17(96), 147-182.
David, Paul A. (1995): The Dynamo and the computer: a historical perspective on the modern productivity paradox. In: American Economic Review, 80, 355.
Donner, Jonathan (2008): Research Approaches to Mobile Use in the Developing World: A Review of the Literature. In: The Information Society, 24(3), 140-159.
Eco, Umberto (1964): Apocalittici o integrati. Milano: Bompiani.
Fortunati, Leopoldina (1998a): Introduction. In: Fortunati, Leopoldina (ed.): Telecomunicando in Europa. Milano: Angeli.
Fortunati, Leopoldina (1998b): "Revêtir des technologies". In: Réseaux, 90, 85-91.
Fortunati, Leopoldina (2005a): The mobile phone as a technological artefact. In: Glotz, Peter / Bertschi, Stefan / Locke, Chris (eds.): Thumb Culture. The Meaning of Mobile Phones for Society. Bielefeld: Transcript Verlag.
Fortunati, Leopoldina (2005b): Mobile Phones and Fashion in Post-Modernity. In: "Future Mobile Phones", Telektronikk, 3_4, 35-48.

Fortunati, Leopoldina (2006a): Mobile communication and its payment. In: Asian communication research, March (Vol.3 No.1), 21-47.

Fortunati, Leopoldina (2006b): User Design and the Democratization of the Mobile Phone. In: First Monday, special issue number 7, September.

Fortunati, Leopoldina (2007a): Understanding Mobile Phone Design. In: Pertierra, Raul (ed.): The Social Construction and Usage of Communication Technologies. Asian and European Experiences. Diliman, Quezon City: The University of the Philippines Press.

Fortunati, Leopoldina (2007b): Mobilnik kot cetrta komunikacijska revolucija. In: Vehovar, Vasja (ed.): Mobilne refleksije. Ljubljana: CIP, 9-28.

Fortunati, Leopoldina (2007c): Immaterial Labor and its Machinization, Ephemera. In: Theory & Politics in Organization 7(1), 139-157.

Fortunati, Leopoldina (2008): Mobile Convergence. In: Nyiri, Kristóf (ed.): Integration and Ubiquity. Towards a Philosophy of Telecommunications Convergence. Wien: Passagen Verlag.

Fortunati, Leopoldina (2009): Gender and the Mobile Phone. In: Goggin, Gerard / Hjorth, Larissa (eds.): Mobile Technologies: From Telecommunications to Media. London/New York: Routledge, 23-36.

Fortunati, Leopoldina (forthcoming): Mobile Phone and the Fourth Communicative Revolution.

Fortunati, Leopoldina / Contarello, Alberta (2002): Internet-Mobile convergence: Via similarity or complementarity? Trends in Communication, 9, 81-98.

Fortunati, Leopoldina / Manganelli, Anna Maria (2003): Young people and the mobile telephone. In: Revista de Estudios de Juventud, 57, 57-78.

Fortunati, Leopoldina / Katz, James E. / Riccini, Raimonda (eds.) (2003): Mediating the Human Body: Technology, Communication and Fashion. Erlbaum: Mahwah, New Jersey.

Fortunati, Leopoldina et al. (eds.) (2010): Interaction in Broadband Society. Berlin: Peter Lang.

Gebhardt, Julian (2010): Alfred Schütz and the Media: The inter-subjective constitution of mediated interpersonal communication in everyday life. In: Fortunati, Leopoldina / Vincent, Jane / Gebhardt, Julian / Petrovčič, Andraz / Vershinskaya, Olga (eds.): Interaction in Broadband Society. Berlin: Peter Lang.

Gebhardt, Julian / Greif, Hajo / Raycheva, Lilia / Lobet-Maris, Claire / Lasén, Amparo (eds.) (2010): Experiencing Broadband Society. Berlin: Peter Lang.

Gerbner, George (1983): Ferment in the Field. Special issue of Journal of Communication 33(3), Summer.

Goggin, Gerard (2006): Cell-Phone Culture: Mobile Technology in Everyday Life. London: Routledge.

Goggin, Gerard (ed.) (2008): Mobile Phone Cultures. London: Routledge.
Goggin, Gerard / Hjorth, Larissa (eds.) (2009): Mobile Technologies. From Telecommunications to Media. New York: Routledge.
Haddon, Leslie / Mante, Enid / Sapio, Bartolomeo / Kommonen, Kari-Hans, Fortunati, Leopoldina / Kant, Annevi (eds.) (2005): Everyday Innovators. Researching the Role of Users in Shaping ICTs. Dordrect: Springer.
Höflich, Joachim R. (2009): Mobile Phone Calls and Emotional Stress. In: Vincent, Jane / Fortunati, Leopoldina (eds.): Electronic Emotions. The Mediation of Emotion via Information and Communication Technologies. Oxford: Peter Lang, 63-83.
Höflich, Joachim R. / Gebhardt, Julian (eds.) (2005): Mobile Kommunikation. Perspektiven und Forschungsfelder. Frankfurt am Main: Peter Lang.
Höflich, Joachim R. / Hartmann, Maren (eds.) (2006): An Ethnographic View on Mobile Communication. Berlin: Frank & Timme.
Hughes, Thomas P. (1983): Networks of Power: Electrification in Western Society, 1980-1930. Baltimore, MD: Johns Hopkins University Press.
Jenkins, Henry (2006): Convergence Culture: Where Old and New Media Collide. New York: New York University Press.
Lasén, Amparo (2010): Mobile culture and subjectivities: an example of the shared agency between people and technology. In: Fortunati, Leopoldina et al. (eds.): Interaction in Broadband Society. Berlin: Peter Lang.
Lievrouw, Leah A. / Livingstone, Sonia (eds.) (2002, new edition 2006): The Handbook of New Media. London: Sage.
Kim, Pyungho / Sawhney, Harmeet (2002): A machine-like new medium – theoretical examination of interactive TV. In: Media, Culture & Society 24, 217-233.
Ling, Richard (2003): Fashion and Vulgarity in the Adoption of the Mobile Telephone Among Teens in Norway. In: Fortunati, Leopoldina / Katz, James E. / Riccini, Raimonda (eds.): Mediating the Human Body: Technology, Communication and Fashion. Erlbaum: Mahwah, New Jersey.
Ling, Richard / Pedersen, Per E. (eds.) (2005): Mobile communications. Re-negotiation of the Social Sphere. London: Springer.
McKenzie, Donald / Wajcman, Judy (eds.) (1985): The Social Shaping of Technology. Milton Keynes: Open University Press.
Manovich, Lev (2001): The Language of New Media. Cambridge, MA.: MIT.
Merton, Robert K. (1949): Social Theory and Social Structure. Glencoe: The Free Press.
Nyiri, Kristóf (ed.) (2008): Integration and Ubiquity. Towards a Philosophy of Telecommunications Convergence. Wien: Passagen Verlag.
Oudshoorn, Nelly / Pinch Trevor (eds) (2003): How Users Matter. The Co-Construction of Users and Technologies. Cambridge, MA.: MIT Press.

Oudshoorn, Nelly / Pinch Trevor (2008): User-Technology Relationships: Some Recent Developments. In: Hackett, Edward J. / Amsterdamska, Olga / Lynch, Michael / Wajcman, Judy (eds.): The Handbook of Science and Technology Studies. Cambridge, MA: MIT Press.

Pertierra, Raul / Ugarte, Eduardo F. / Pingol, Alicia / Hernandez, Joel / Dacanay, Nikos Lexis (2002): Txt-ing Selves: Cellphones and Philippine Modernity. Manila: De La Salle University Press.

Pertierra, Raul (ed.) (2007): The Social Construction and Usage of Communication Technologies. Asian and European Experiences. Diliman, Quezon City: The University of the Philippines Press.

Pool, Ithiel de Sola (1983): Technologies of Freedom. Cambridge, MA: MIT Press.

Rice, Ronald E. et al. (eds.) (1984): New media: Communication, Research and Technology. Beverly Hills, CA: Sage.

Ropohl, Günter (1999): Philosophy of Socio-Technical Systems, Society for Philosophy and Technology, 4(3). URL: http://scholar.lib.vt.edu/ejournals /SPT/v4_n3html/ROPOHL.html#emery.

Schramm, Wilbur (1977): Big media, little media: Tools and technologies for instruction. Beverly Hills, Calif.: Sage.

Schutz, Alfred (1967): Phenomenology of the Social World, Evanston, III: Northwestern University Press.

Schütz, Alfred / Luckmann, Thomas (2003): Strukturen der Lebenswelt. Frankfurt am Main: Suhrkamp.

Star, Susan Leigh / Bowker, Geoffrey C. (2006): The infrastructure of new media. In: Lievrouw, Leah / Livingstone, Sonia (eds.): The Handbook of New Media. London: Sage.

Van Dijk, Jan (1999): The Network Society. An Introduction to the Social Aspects of New Media. London: Sage.

Vincent, Jane (2003): Emotions and Mobile Phones. In: Nyiri, Kristóf (ed.): Mobile Democracy. Essays on Society, Self and Politics. Vienna: Passagen Verlag.

Vincent, Jane / Fortunati, Leopoldina (eds.) (2009): Electronic Emotion. The Mediation of Emotion via Information and Communication Technologies. Oxford: Peter Lang

Virilio, Paul (1992): Big Optics. In: Weibel, Peter (ed.): On justifying the hypothetical nature of art and the nonidenticality within the object world. Köln: König.

Williams, Raymond (1975): Television: Technology and Cultural Form. New York: Schocken Books.

Williams, Frederick (1982): The Communications Revolution. Beverly Hills, CA: Sage.

Whittaker, Steve (n. d.): Theories and Methods in Mediated Communication. URL: dis.shef.ac.uk/stevewhittaker/CMC-review02.pdf.

Yung, Vicki (2005): The Construction of Symbolic Values of the Mobile Phone in the Hong Kong Chinese Print Media. In: Ling, Richard / Pedersen, Per E. (eds.): Mobile communications. Re-negotiation of the Social Sphere. London: Springer.

Absence to presence: a vision of the communicating human in computer-mediated communications technology research

Richard Harper

Introduction

In 1904, Edouard Estaunie coined the term telecommunication by merging the Latin word communico, to impart or share, with the Greek word tele, for distance. He had in mind a word for any technology that used electronic signals to exchange information. We still use the term in pretty much the same way one hundred years later, though in a more encompassing fashion: the historian of telecommunication, A. A. Huurdeman (2003), for example, uses it as a label for any 'technology of information transport' – one assumes that papermail would fit into this category, as well as motorcycle couriers. Somehow, though, we don't think of telecommunication as quite the word for our communications obsessed times. One might say that the word computer and more especially the words the internet do instead; these words have become rough and ready synonyms for the all too human desire to be in touch. What I am thinking of here is not the evolution of words but simply the use of everyday phrases like "Are you on the internet?" and, "I am so much better at keeping in touch with a PC". What these phrases convey is that some 'thing', a PC, a mobile, 'the internet' and so on – whatever it might be – is like an umbilical cord linking each and every one of us in principle; an umbilical cord between ourselves and our friends, ourselves and the world at large.

In this paper I am not interested in the words we use but I am interested in trying to define the kinds of vision of the human, the user, the one desiring to be in touch that underscores the inventive landscape I see around me. Key to this vision has been, I think, the idea that more is better, that we should keep our inventive imagination to the mill stone so as to produce more ways of being in touch. It is this, I think, that has produced a landscape where we don't just talk about being in touch via computers as if it were commonplace but expect, also, that this landscape will keep changing, developing and altering as new tools for 'richer', 'more' and hence 'better' communication appear. I won't ask in this paper whether we need more communications channels (though I do elsewhere) but I will try and answer why it is we keep inventing them by claiming that it is an idea about the human – about what the human is – that drives this invention and inventiveness.

I want to do so by recounting and describing some of the thinking of my own world, the world I have lived in for twenty years or so. When I say my world I mean, in particular, a set of places that have been concerned with inventing just the technologies that are of interest to us now. At first glance, this world might not seem the one that has been driving the communications landscape around us. I currently work in Microsoft Research. Now, though this place will obviously have had a great influence on what one sees on most PCs, it might not be thought that it has also helped create the mobile world that we now live in. Some time before now I worked in Xerox PARC's European lab, EuroPARC. Here too one might imagine that the research carried out was about office life and not about being in touch. But I want to say that in practice these places have been very much in the business of inventing the current landscape. This is because these landscapes were being constructed, it seems to me, by what one might call a passion for computer-mediated communication. This passion has been bound up with the emergence of the Web and, most especially, the underscoring set of protocols that allow data traffic to move around the world at enormous speed and in huge volumes[1].

I will want to characterise what I will call a model of the 'interacting human' we have in mind when we invent, test, and explore. This model, I will suggest, has been essentially Turing theoretic. When I say this, the view I will be putting forward may not be the one the reader expects. I will not be arguing about a vision of the human that has to do with intelligence, for example. After all Turing is famous for his so-called test of the same. This entailed asking a machine the same question as a human and if the machine did as well, or its answers could not be distinguished from those of the human, then it would be deemed as having passed the test and ought to be treated as 'equal' to (or indistinguishable from) a human. The question of intelligence is largely immaterial to me; though what is of relevance is how Turing came to answer it himself. For he encouraged a view of the human that essentially turned around a dichotomy, one between the body and the mind, and moreover encouraged most concern with only one of this pair, namely the body. In his view (certainly in his early views if not the ones he developed later on), it was the body that could be seen as delivering the 'intelligent act', whether that body was made of skin and bone or metal and plastic. What went on inside the other half of his dualism, in the 'mind', didn't matter. I will want to suggest that in so doing Turing fostered

1 As it happens, the mobile phone and network manufacturers too have research labs in this space. Mobile phones have become small, handheld computers, not a different technology species from PCs altogether as they used to be; similarly the networks that support them are essentially vast, air-based ethernets. Needless to say, such places, whatever their organisational provenance, are not the only drivers of change and moreover, much of what they invent turns out to miss the mark, if by measures of success we use take-up by the mass public. As we shall see, much of what my colleagues and I invent goes nowhere.

what one might say is almost a whimsical indifference to important properties of human action, of human communication in particular. When I say whimsical I mean a casualness and lightness of thought – he didn't worry too much about what the mind as linguistic category might do or mean, and even less about its correlate, 'mindfulness'. As I shall show, this neglect is one that his followers or at least those trained in his 'science' (computer science) rarely think about carefully, just as the possibility that the dualism itself might be egregious is hardly ever countenanced too. When I say thought about carefully I am meaning by those who are in the inventing business. I won't be getting into arcane, scholarly arguments about the specifics of what Turing said, nor will I want to report on the enormous literature on the history of ideas related to his work. What I am interested in is the pragmatic use of his thinking (albeit somewhat simplified), a view that is manifest, I will want to say, in the inventiveness I see around me, some seventy years later. I see colleagues who are concerned with what one might call 'the doings of the body' and not with 'the doings of the mind'.

One consequence of this might seem a philosophical aside, and that is that this view treats human behaviour much like the behaviour of any animal, and not especially or uniquely human – it's all about body movement and nothing else. I think this is profoundly egregious, and can lead one to misunderstand what is human about communication between humans. A second consequence, more concrete, is that this vision leads to all sorts of paradoxes and ironies as 'users' of the technologies (and by users I will be including myself and my colleagues) when we are confronted with things (of our own devising!) invented for the body (so to speak) into doing things that are for neither the body nor the mind but for a world in which the two aren't separated. This is especially so when communication between persons is at issue. As we shall see, it is a peculiar endeavour to separate the mind and the body when it comes to designing technologies of human communication since the thing that is the main material of that communication, expressions, intentions, aren't best thought of either in terms of the body or the mind: they are expressions and intentions, not body movements. Nor are the movements that embody expressions and intentions proxies for 'something' going on 'somewhere in the mind' (i.e. external representations of some internal act).

Interacting individuals

I want to start my inquiries with a vignette of things I was involved with some twenty years ago. I want to describe two or three systems that we were toying with then, say something about our expectations as regards what they let us do and then say something about their fate. I then turn to describe how I still see colleagues, today, playing with variants of some of these concepts (by concepts I mean a kind of invention). I won't want to say that here is proof that my colleagues and I don't seem to get anywhere but I will want to say that what motivated us then is still pretty much what motivates us now.

What motivated us then had to do with mix of hope and expectation obviously. But beyond this what motivated us then was a notion about what a human is, what an interacting human does, and thus, with that in mind, what our inventiveness ought to enable. We are entranced by the same view today.

So, to begin: in the early Eighties, I was fortunate enough to work at Xerox EuroPARC, also here in Cambridge. EuroPARC was a curious place to work, in lots of ways, and one was the predilection of many of its staff for video-mediated communication or VMC. One consequence of this was that the EuroPARC building – the entire building – was wired for sight and sound. More or less each and every office had a coaxial cable linking it to every other (there was a lot of cabling!) and these conveyed high quality audio-video connections between offices. Thus what was called a 'media space' between any two or more offices could be created. Part of the fun of working at EuroPARC was the shock and glee that the media space provoked. Some journalists were appalled, thinking it Big Brotherish; others enamoured at how the media space was pointing towards the prospect of being able, in the future, to bring people separated by great distances, closer together within their office worlds. Needless to say, many projects were undertaken investigating the utility of the system - or rather set of systems – that comprised the computational infrastructure of EuroPARC's media space. Now, looking back, a couple are of note.

The first had to do with shared editors or what were most often called shared whiteboards. These allowed people in different offices to type text into a document at the same time as someone else, elsewhere, in another office. These shared editors worked as follows: one of us would type text into a window on a screen (the 'editor window' of the application); our colleagues in another room would see this being typed into their view on the editor window and would then, if they so desired, type on the next line of that window. This would be visible to the person who created the first line. Typically, one could not tell who wrote what unless one of us typed their name at the end or – and this was more common – one could tell simply by the turn-taking. I wrote, she wrote, I replied; that was the gist of it.

Of course, our actual use was a bit more complex than this. Also, the reader may get suspicious that they are familiar with contemporary applications that sound very similar: they will be right. But before I come to that let me explain some more about what we wanted with these editors. These applications were being investigated for their use in what was called knowledge work. Whatever knowledge might be – it's a rather vague term at the best of times – we were interested especially in supporting such work when the participants were separated by distance. The thought was that new organisational structures were increasing the likelihood that such work would be undertaken by individuals spread around the world. Organisations were going global; knowledge work was being done in New York, London and New Delhi; technologies to support distributed knowledge work were required. Though our office was only small while the distances of the global organisation enormous, it was thought that the distance between our individual offices could be treated, analytically, as proxy for bigger ones in the real world.

So, enough of why we thought we were looking at knowledge work. Now back to the technologies: the whiteboards. Nothing much came of them, whatever the merits of investigating knowledge work. By nothing much I mean we did not manage to implement them in any effective way in 'real work', that is to say in work other than our own. When I say manage, I am perhaps being disingenuous. Though these tools clearly were central to what one might have said was our research remit, we did not find them hugely interesting. They seemed altogether too prosaic to be worth investigating (though I should say some visiting researchers did find interesting aspects to them which they published in the groupware and CSCW literature). Our managers remained vaguely hopeful that we might evolve these shared editors in novel ways, bringing a Xerox sheen to what was a fairly crude technology. But we did not. Other technologies, especially the audio visual aspects of the media space, enticed us away.

Before I say why, let me note that nevertheless we used these shared editors for ourselves. But we did not use them for knowledge work – to help us write reports or share profound ideas. We used them to play. We used them because they became devices for laughter and mischief; because using them could be fun. The topics were mostly jocular, asides about the burden of deadlines, occasionally about plans for after-work activities. At least one romance flourished at EuroPARC with this form of exchange. I am mentioning this now because what we were playing with then was the genesis – albeit in a convoluted and indirect way – of what we now know as instant messaging – as the reader may have recognised. What our managers thought would be tools for professionals and their knowledge practices, indeed what my research colleagues and I dismissively assumed would be work-orientated applications, ended up being crafted to support, some twenty years later, such things as teenagers keeping in touch with each other when they have been 'imprisoned' in their bedrooms by parents hopeful that once there, they might then do some homework. Of course, I am making fun of instant messaging and the wiles of teenagers (cf. for example Harper 2005a, 2005b; Harper / Hamill 2005). But my point is a serious one. Tools for serious things like work have ended up being used as tools for play in social life.

Why we believed that these applications were to be thought of only in terms of work seems a little perplexing now. After all, we did not use them to support our own work even though we were professionals doing knowledge work of a kind. As I say, we found fun and utility in them through their playfulness, in supporting not professional engagement but our social selves – we were doing then pretty much what teenagers do now. One might pursue this line of argument and suggest that we were indulging in what the French sociologist Maffesoli suggests, metaphorically, in his The Time of the Tribes (1996), is a contemporary form of tribalism or more specifically, a Dionysian tribalism. He has in mind the desire of our 'modern selves' to celebrate the temporary, daily, bodily togetherness of work tribes, whichever tribe that might be – research lab tribes included (cf. my own such study, Harper 1992). I think there is certainly something in this metaphor, but it is not appropriate to develop it here (cf. Harper 1996).

Who is the user that is invented for?

Going back to my concern, with the 'kind of human' underscoring our inventiveness, the fate of this technology can lead one to ask a number of questions. One might ask, for instance, whether there is something peculiar in the relationship between those who do the inventing and those for whom the inventing is said to be done. Is this an instance of inventors justifying their ideas on the grounds that someone, if not themselves, would use their inventions when it turns out these 'some ones', these 'ordinary people in the real world', are something of a construct; even artificial, one might say? This leads one to think of the term Otherness used to label the concern of anthropologists: one might say that one of the ways that we work in my world is by making users 'other' in various ways and here, this otherness was simply other than ourselves. Certainly I would be the first to admit that, over the years I have indeed heard my colleagues justify some idea(s) with the assertion that 'real people' will use them, even if the colleagues in question (the inventors of the thing) don't say who these people might be. It is probably true to say, also, that my colleagues, just like me, can drift into laziness; tiredness even and so don't really have in mind a very particular 'real world user'. And besides, phrases like 'ordinary people', 'real people' and 'people in the real world' do sometimes get used as a kind of bland catch-all to justify technologies that my colleagues and I haven't really figured out the purposes of ourselves. We think we have devised something that is of use, but for what and for whom is beyond our comprehension. Ordinary people have lots of faces and lots of desires and so we use this label as an expression of hope, if nothing else, a hope that some one, some where, for some reason or other, will indeed like our contraption(s).

Such reflections, even if they are accurate on occasion, are a little unkind. In some instances we don't have a clue who might use our inventions, but this is not very common. By and large my colleagues and I are pretty thorough in our work and most often do indeed have a view of the 'user' in mind. This view, earnestly held, seriously thought about, is not, I suggest, a view of the user that one might hold outside of the work places I am describing. I want to suggest that this view, not always articulated in-depth it has to be said, is nevertheless the view that the majority of my colleagues assume is one that they can 'work on'. They held it then; they hold it now. These examples of shared editors have only hinted at what this view is. Let me now uncover it a little more clearly by recounting another technological experience afforded by the EuroPARC media space. Our absorption and fascination for this will tell us a great deal, I think, about the kind of user we had in mind. As we get a better sense of it, we will also begin to see, I hope, how this vision motivates us today.

Ways of 'looking' in my world

At the same time as we were using these shared editors, another project was seeking to add some refinements to the media space, especially the audio video conferencing experience afforded by it. When I had first arrived at EuroPARC, the media space system was fairly simple. Each office had a camera – a high quality camera – mounted on a wall or tripod, or attached to some shelves. It was through this that others, using the media space, could see you. There were speakers, too, for sound connection and these would be fitted in the room, but not necessarily beside the screen; they would be put anywhere space could be found. There would be a monitor as well which let one see the 'other' office when a connection was made, and these two might be elsewhere in the room. All these devices (the monitor, camera, speakers) were separate from the workstations we were using – though we used the workstation to initiate a 'connection'. One consequence of this arrangement was that if one were to look into the camera then one might well be looking away from the monitor on which one could see one's colleague. The same would hold true for one's colleagues. Hence, when one made a connection to another, one would look at them on one's own monitor but they would appear to be looking elsewhere – not at us. They too were looking at their monitor wondering why we were looking away. In both cases, the camera would be some way from the monitor – perhaps not far – but far enough for this 'incongruency of perspective' to occur (as my colleagues expressed it). Some of us found this vexing or at least a cause of mirth. 'Look at me when I am talking!' they would shout; 'I am!' we would reply. Some of my colleagues, though, were convinced that this was problem that needed fixing. They devised solutions, inventions of sorts, that solved the problem. Doing so was not difficult – somehow the cameras needed to look at the subjects as if the cameras were looking from the same point of view as the monitor. Both the looking and the glancing needed to be the same.

At that time, in the early Eighties, the only way of achieving this 'fix' was to use a system of mirrors, like a periscope, assembled and boxed up around the monitors such that the line of sight of the camera appeared to be hidden within the monitor. The system – or rather this improvement to the system – was called video tunnels. Once built, various experiments were undertaken to investigate whether 'users' (us) felt more at ease with the system. One particular concern was to see whether users felt better able to understand what others in the system were looking at. With this system, one knew that when someone looked away they weren't looking at the picture of oneself, for example, and therefore, might have demonstrable purposes with some object, a document say, that one might orient to oneself. There might be a 'mutual referent' as it was grandly put. These findings were published. As it happens, once the experiments were completed, the system was packed up. The periscope type boxing was somewhat large and unsightly; besides none of us seemed to care for the difference video tunnels made. Within weeks the set up was dismantled and the media space returned to its former state.

Be that as it may, from my own view, it certainly seemed easy to use a system when the eyes of the fellow participant seemed to be looking at oneself, but at the same time much of my own enjoyment with the media space had nothing to do with the ability to look it another's eyes as they gazed into one's own. It was fun merely to peer into another room, to see what was happening there. Sometimes we would use the media space to simply call out, between rooms separated by several floors, and see if someone was around. This was especially so between reception and the admin offices, on the bottom and top floors respectively. This play did indeed seem a kind of Dionysian tribalism, as Maffesoli would have it; all the more so as I look back now and wonder at how much I played when I was younger.

I am not wanting to comment on that playful tribalism, though. As I say, I am interested in the view of the user that motivated this enhancement to the media space. I want to get a little clearer on that view by recounting how, now, many years later, my colleagues, here, at another research lab, albeit in the same city, are endeavouring to invent something very similar. But this time, they are devising something without the ugly boxing, the heavy camera or the co-axial cables. Whereas then we had to make do with very large cameras, now video cameras can be very small indeed. Moreover, then the best we could do was stream the video data over the network, now we can process it and interrogate it. This makes a considerable difference to what can be assembled, or as I am wanting to call it, invented.

The new system I have in mind combines two very small video cameras attached to the top of the monitors at either end of the link (in two offices, say). As before, the cameras do the looking, as it were, while the monitor's display what is to be seen. Thus far this new system and the one in the media space are the same, albeit in size and numbers of cameras. But here the similarities end. In the new system, data from the cameras are processed with so-called intelligent algorithms so that the system can 'recognise' and follow the movements of those looking at the monitor. In simple terms the cameras produce stereoscopic data that the computer can analyse so as to identify where the user's head is, so to speak. There are various ways of doing this, from tracking movement of the head itself (as a three dimensional outline) through to tracking of specific points on the head, such as the eyes, tip of the nose and mouth. Irrespective of the tools (or 'keys' if you prefer) the system uses, the result of this arrangement is that, once the tracking is processed, the computer can adjust the image shown on the remote monitor to ensure that when the user is gazing at their own monitor, it looks as if they are gazing through the cameras at the remote viewer. The actual distance between the cameras and the screen is in effect dissolved. The end result of this system is quite delightful, it has to be said, albeit slightly odd. Odd because one knows that the remote person's eyes are not really gazing at you; they are looking at a screen displaying a virtual you. But delightful too since it affords a kind of playful paradox: it's nice to be looked at, but after a while one tries to see how far one has to go before one cannot be looked at – i.e. to go beyond the processing of the system so that it cannot 'correct'.

What was the goal behind these two systems? In one way I have said it already and it seems obvious, indeed, perfectly reasonable. When one is having a video conference (or videocall), it would be nicer if the person you were conferencing with appeared to be looking at you, rather than somewhere else. Nicer seems a weak word here, but it does convey the sense of greater politeness and social grace that the system afforded. The point I am wanting to make though has to do with the relationship between my colleagues ideas of what to design for, ideas related to what the human actor or user of the system was – and how to deliver, through design, this sense of grace. It is certainly true to say that people, when together or when using a conferencing system, will somehow manage to co-ordinate their gazes and orient their behaviours in such a fashion that they produce for each other a sense that they are 'in touch', 'of a mind', and 'doing something collaboratively'. It is certainly true also that there is a great deal of empirical literature on how this sense of 'jointness' is achieved, especially in the ethnomethodological cannon (cf. for example Sudnow 1972). But somehow our research wasn't really about how this synchrony of looking would make for courtesy. Systems of social grace and propriety were not the motive behind the design of the system. Rather, what was behind the system was what one might call an idea about the body mechanics of interacting humans. I am thinking here of a design which sought to allow lookings and glancings 'fit together'.

I need to be careful here because there is an extensive literature reporting on the moral dimensions of looking and glancing – researchers like Egon Bittner (1977) have explored how not responding to a look can be seen as social rebuff, for example, just like not answering a hello. His paper 'Must We Say What We Mean?' points out that meaning is conveyed even when we don't speak or even don't look. This is because of how our actions embody our intentions and are thus visible for others to see. David Sudnow (1972: 259-280) in the book mentioned above does the same. But this argument (that meaning is incarnate in our actions), though I think it is right and I will come back to it in later chapters, is not what I am wanting pursue now. What I am wanting to suggest is that what we devised was a much simpler vision of human interaction or interacting humans that somehow emphasised just the body mechanics of the interaction that somehow lost sight of this moral dimension.

Fitting touches

Why this should concern me? Why is this significant? Well for one reason, I am trying to get to the view of the human that motivates us and these examples are of systems developed some years ago. I now want to refer to another system my colleagues have been developing more recently, also a conference system of sorts, and show that this has a similar concept of 'body mechanics' as the key of human communication behind it. This view is persistent, I want to say. If the systems I have mentioned thus far have concerned themselves with the 'how' of looking faces, the system I now want to describe concerns itself with the touching and moving of hands.

The system I have in mind uses a stereo camera and some clever object recognition software to notice when hands are placed on a digital document, when they point, for example, or even when they appear to erase a word. The cleverness, the inventiveness, is making such movements recognisable to the system. It is made possible by developing, first, the ability of the machine to discern shapes, in particular the shape of a hand and a finger in this case, from other shapes and forms that the cameras might see. Second, this is combined with the capacity to determine movements of these things when the movement in question indicates something. Hence, a finger's movement can be seen as a gesture, for example, not a mere shadow moving over a document. The result is a system where users can get to grips with their digital documents. They can use their hands, moving on top of a document, to mark out, edit, remove, paste, as they so desire, words, images, stuff if you like, in the document. The movements of their hands, the gestures of their fingers, and so on, are converted by the machine into what one might think of as equivalents to mouse pointings and clickings. Such a technique is not devised so as to replace mouse and pointing, as it is a method for interaction over documents that are being interacted with by people over distance. One person points and shakes his finger over a document; a person somewhere else can see this interaction and can have the consequence of this interaction made manifest in the document they are working on. This is not merely, then, a picture of a hand on a document, it is a hand interacting with a document that itself is digital; and a document which is itself shared.

What I would like to suggest is that with this and the prior examples, one can begin to see what my colleagues have in mind as regards how to conceive of the user. We have in mind the idea that, whatever it is that people want to do, they can be aided in that task by systems that replicate what one might call the physical geography of the interactions in question. In the first case, in the media space examples, we imagined that, if we were designing a communication experience for people who want to communicate to those far away, then one way of supporting that would be to invent a system that would let the people in question glance and look and gaze at each other when they are remote just as they might when they were together. What we needed to design for was what one might call the interactional geography of lookings and glancings. In the other example of the more recent system, we were also concerned with remote communication, but in this case, interaction and communication about documents. Here a system was invented that meant that how someone uses their hands when dealing with a document could be conveyed, replicated to someone remotely. Here it is the geography of touch that mattered.

The way I have put it, as geographies of touch and lookings, makes our vision of the human and of human needs in communication technologies seem obscure, even arcane. This is not my purpose. My mannered description is designed to make strange what is in fact commonplace in my world, the world of inventing computer mediated communications systems. Making it strange can help us think about it, I hope. What I am suggesting is that what motivates a great deal of the inventiveness I see around me, as illustrated with the enhancements to the systems mentioned above, is a concern with bodies, with the human body, or more particularly, with the bodily mechanics of human interaction with computers and hence, also communication through and with computers to other people. I am saying that inventiveness in my world is about devising ways of allowing the mechanics of the communicating individual to fit together. In this vision, it's not what is said in communication that matters, nor why something might be said, nor what may be the consequences of the saying of something. It is in the how of the saying, the manifest behaviour entailed when that thing is said.

When using the word 'said' I am obviously evoking words, but as the examples above should attest, in this vision, it is not just words that matter.

What it is that is exchanged in a communication might be broad indeed – glancings, pointings or gestures, just as much as the sharing of the written or the spoken word. In all cases, I am saying that I and my colleagues sought then and seek now ways of letting the body mechanics of those doings be replicated. Replicated so that they are conveyed over distance, for example; replicated so that the assembly of those mechanics, the sounds, the glancings, the pointings, may be conveyed as an 'assembly'. And I am thinking too, of how the interaction of multiple, communicating bodies needs to be replicated – glancings need to be synchronised; listenings need to be two way; distractions ought to be shared if they are to be oriented by another body's lookings and glancings.

A vision of the human user

Now, a number of questions and queries naturally follow on from this. One obvious question is why this view? Why not another? How does this view affect or drive invention? Will there come a time when bodily communicative acts can be entirely replicated, for example? This seems unlikely, but what are the difficulties here? Besides, the examples I have used seem rather confined, essentially to do with knowledge work and conferencing of various kinds – so what about the daily assault of messaging and callings and blog postings we see around us today? These don't seem like body mechanics; so where have these come from – what inventive landscape produced them? Aside from these matters, what about motivating inventiveness: is satisfying this 'ideal' (of bodily fittings, irrespective of whether it is right or wrong) motive enough to keep my colleagues and I at the mill stone? Could other visions of the human perform a similar role: namely the role of keeping researchers passionate enough to keep on inventing 'this' and inventing 'that'?

One can't answer all of these questions, though some are certainly more important to our concerns. Let me deal with the question of where this view of the human comes from; this can provide a basis for commenting on other questions like motivation. It would appear that my colleagues and I have a vision of the human, the user if you will, and in particular a vision of the actions of the human, which one might say is not really very common – ordinarily we don't think of fitting our bodies together when we communicate (or rather sometimes we do but not often). Though, as I said earlier, it is sometimes the case that my colleagues coin the phrase 'ordinary user', it seems to me it is not a very ordinary vision of people they have in mind.

Following on from the formulation that Peters (1996) uses in his 'Speaking into the Air', I think there is a metaphysics here, a set of ideas about who and what the human communicator is. In this case these derive from various sources within computer science, and I think one of the most influential was a founding father of this discipline: Alan Turing (1936). I suggested at the outset that his influence relates to his encouragement of a view that splits the human into either body or mind, and tends to encourage the disregarding of the latter or at best a whimsical approach to it. As it happens, this might not be a precise representation of Turing himself, of the detailed claims he made in his various papers – there is no mention of bodyism in his work for example – but it is I think a fair comment of the consequences of his view in the processes I am thinking of – inventive ones, ones concerned with technologies for human communication.

More particularly, in his various papers on 'intelligence' and the 'computer machine' written in the Thirties and Forties, Turing claimed that one may understand intelligence by treating it as solely to be measured in external performance, in behaviour (cf. for example his 1936 paper, 'On computable numbers, with an application to the Entscheidungsproblem', loosely meaning the 'halting problem'). Whether a machine or a person performs that behaviour was irrelevant, he proposed. This had methodological advantages for him, offering, as he saw it, a route around the problem of moral judgement. A person could not be relied upon to objectively describe a person's action as intelligent or otherwise for fear of upsetting the person being judged. Thus, in Turing's perspective, only behaviour must count and the source of that behaviour must be hidden from view. The fact that in so doing Turing reduced the concept of intelligence from being rich, complex and subtle, used in diverse ways to achieve different sorts of understandings, purposes and descriptive values in ordinary language and life, into something that is little more than a label for a binary opposition, where an action is either intelligent or not, seems not to have worried him.

I don't want to pause and comment on that at any length, here, though. Of more importance is that this move has another consequence, as follows. Those who adopt Turing's view assume that what goes on inside the machine itself is not only invisible, it is somehow tricky and best avoided. The 'moral implications of judgement' or, rather, the moral-aspects-of-a-mind's-thoughts are such that computer scientists ought to steer clear of them if they are to do 'science' (as Turing conceived of it). Now, as it happens, this hasn't meant that computers scientists subsequently haven't looked at what goes on 'inside the head' and attempted to do a kind of science (one that Turing would approve of). But when they have, they have done so from the presumption that one starts with bodily behaviour, with body-ness if you will, and have to construct, with external evidence, ideas about what goes on inside the head. For some of those who have taken this Turing theoretic view seriously, the implication is the possibility that if a machine, a computer machine, can equal a human through the use of a programme (i.e. some lines of code as in a Turing test), then perhaps it is the case that the human mind is computer-like too. In their opinion, one should research to discover what that code that makes-the-mind-work might be. In computer science this view is held by the likes of Marr (1983), who claimed that the way a computer machine can recognise a visual object (via various forms of Bayesian prediction, for example) must be the way a human mind processes and hence 'sees'. Outside of computer science, this idea has been adopted too, with much of contemporary cognitive science being a product of the possibility that one can divide the human into the body and mind and assume, also, that the mind is a processing machine, just like a computer (For a review cf. Sheutz 2002).

One could go on. Turing has been enormously influential in all sorts of ways. But here I am not wanting to offer a general history: I could explore how Turing's view can be seen to have its roots in earlier ideas – Cartesian dualism, and even before that, to when the mechanism of clocks provided inspiration for how to grasp what the human body was (cf. Hacker 2007: 233-257). But that is not my concern. I am wanting to recount the inventiveness I have seen around me in terms of a vision of the human that has underscored it. I am suggesting that a concern with body mechanics has motivated us and I claim this not because I think we have adopted that concern out of intellectual investigation into philosophical ideas about 'humanness'. If, as the sociologists might have it, we have configured our view of the user, we have done so first and foremost for pragmatic reasons. When I say this I mean that we have sought visions of human endeavour and the human actor performing those endeavours that have prompted ideas that could lead to invention. The body emphasising view I have described has helped us to achieve this. It has highlighted 'problems' and 'needs', as we saw them, the kinds of things we thought we could 'solve'.

In saying this it might appear that I am suggesting we have adopted a poor or enfeebled vision of the human. This may be the case from some points of view. But this is not what I am arguing. The view in question ought to be judged by what it let us do. I want to suggest it lead us to what we thought were complex designs, not simple designs. In these respects, this view worked for us[2]. That is to say, the examples show that the credit one ought to give to the model of the human we used is in terms of what it led us to invent. We thought it would lead us to be focused on more complex technologies than other views might have done. As I say, we turned away from early versions of instant messaging because we thought them technologically prosaic and mundane in terms of the richness of the (human) communication in question. We turned instead to the technologically more complex audio visual systems. As we did so we thought that we were addressing more complex and profound issues of human

2 One might say that Turing's view of the human in the 1936 paper also allowed him to do *his* work, namely to define some parameters of computer algorithms. That was the focus of his thinking, not some ontology of the human. As it happens, in later years Turing did drift in to that concern but that is another matter.

communication. Our view led us to focus on complex bodily matters, not matters that might be measured in terms of human value or intention. As remarked, we used the shared whiteboards for some fairly complex social actions – play, romance – but viewed those behaviours as less interesting to invent for because they used only simple technologies and from our pragmatic standpoint only simple behaviours. In our judgment, they used only one mode of bodily action – namely the keyboard. Our other inventions seemed more appealing since they used multiple modes of bodily action as well as more complex technologies. What appealed to us were precisely more sensual dimensions in communication. This appealed because more dimensions of this kind would demand more innovative technological solutions, we thought.

Conclusion: squeezing humanness into body movement

How does one judge such motivations? Should one compare them against richer visions of the human say, or simply in terms of how many inventions derive? These are not the same measures at all. I think we ought to be kind about these motivations and generous too in our measurement of them. This vision did drive us to invent, as should be clear, this inventive spirit did indeed produce a lot of 'IP' (as the saying has it in corporate life: IP stands for invention proposals: these are the first documented stages of a case of a patent).

Nevertheless, there were paradoxes observable in the circumstances of the work in question just as there was in the scope of the inventive drive. For example (and as I mentioned), when the video tunnels experiments were completed, we packed up the enhanced system and went back to the old audio visual system. Why was that? Of course there were practical reasons: the new set up was bulky for example. But one might also add that we didn't imagine we would find it useful for ourselves, in our own activity. This was because what it offered, a richer set of bodily fittings, did not actually equal an easier mode of expression. Indeed, quite the contrary. When one used the video tunnels or indeed whenever one used the simpler audio visual system, the question of bodily fittings, of mutual glancings and lookings, was only part of consequences or implications of the actions in question. When I shared a video connection with others, those others weren't primarily interested in the particular objects I looked at; they wanted to know why I wanted to glance at them at all. To make a video connection was an enormously powerful act, an assertion of a kind of intimacy, then, a closeness that was not something that each and every one of my colleagues thought 'appropriate'. By appropriate here I am alluding to the moral implications of such acts, not their physiological ones. But such concerns, though 'real' were beyond the scope of our inventive landscape. We did not worry about these concerns. Besides, if we did bother to look at them closer (over coffee, say, or when we reflected at the end of the day) we saw how tricky the concerns in question would be to unpack and use as a basis for design. Sure, we recognised that there were issues, concerns, topics, properties if you like of human action that were relevant here, but our instincts were to avoid them.

If we had problems with fitting the mind's landscape to the audio visual system, to the bodily landscape the system enabled, then we have even more problems fitting the latest systems we have devised to allow interaction over distance. At least with the EuroPARC audio visual system we played with the infrastructure, but with our new interaction-rich conference systems we don't even play. Once experiments are completed they are packed up and put away. None of my colleagues have any desire to use these systems for themselves. Again, they use the phrase 'ordinary people' to suggest that others, somewhere else, might be willing to do so. But why they don't themselves want to use such is something that they view as uninteresting. Answers to that question don't lead them to think of any new inventions. It doesn't seem to have any pragmatic value.

This attitude reflects a whimsical indifference to intention and expression, it seems to me; a casual looking away from what one might label mindfulness in human action. I think Turing is the source of this, as I suggested at the start of the paper. He focused on the bodily aspect of human action and thought little about the mind or, as I am trying to suggest now, more accurately what one would describe without evoking this dualism, as the expressive and intentional, and so too do my colleagues and I now. Of course other things might be done with the technologies we invent, and of course human endeavours encompass more than we designed for. But that 'more', whatever it might entail, didn't – and most often doesn't still – interest us when we invent. We focus instead on what we think is a better root to inventiveness. Never mind that this leads us to disregard technologies that might be more influential in the world at large (such as I.M.); never mind either that the audio-visual technologies that we invent seem to be introduced to a world that seems largely unimpressed. We are impressed by our inventions. We thought – we think – we are inventing for a richer human experience; we think that this will keep us at the cutting edge of invention. That is why we do it I want to say. But what we end up inventing for is a human that seems somehow devoid of the very thing that makes humans human – their intentions, their expressiveness, the thing that makes their action not merely action, but meaningful. In this paper, I have wanted to explain why this seemingly odd consequence of our inventive desires arises: not because we are simple minded, but because our pragmatic visions lead us to it.

References

Bittner, Egon (1977): Must We Say What We Mean? In: Ostwald, Peter F. (ed.): Communication and Social Interaction, New York: Grune & Stratton, 180-197.

Hacker, Peter M.S. (2007): Human Nature: The categorical Framework. Oxford: Blackwell.

Harper, Richard (1992): Looking at Ourselves: An Examination of the Social Organisation of Two Research Laboratories. In: Proceedings of CSCW '92. New York: ACM, 330-337.

Harper, Richard (1996): Why People Do and Don't Wear Active Badges: A Case Study. In: CSCW: An International Journal, Vol. 4, 297-318.

Harper, Richard (2005a): To Teenage Life to Victorian Morals and Back: technological change and teenage life. In: Glotz, Peter / Bertschi, Stefan / Locke, Chris (eds.): Thumb Culture: the Meaning of Mobile Phones for Society. Bielefield: Transcript Verlag, 101-113.

Harper, Richard (2005b): The Moral Order of Text: Explorations in the social performance of SMS. In: Höflich, Joachim R. / Gebhart, Julian (eds.): Mobile Communication – Perspectives and Current Research Fields. Berlin: Peter Lang, 199-222.

Harper, Richard / Hamill, Lynne (2005): Kids will be Kids: the Role of Mobiles in Teenage Life. In: Hamill, Lynne / Lasén, Amparo (eds.): Mobile World Past, Present and Future. London: Springer-Verlag UK, 61-73.

Huurdeman, Anton A. (2003): The Worldwide History of Telecommunications. New Jersey: Wiley.

Maffesoli, Michel (1996): The Time of the Tribes: The Decline of Individualism in Mass Society. London: Sage.

Marr, David (1983): Vision: A computational Investigation into the Human Representation and Processing of Visual Information. New York: Freeman & Co.

Peters, John Durham (1999): Speaking it the Air: A history of the idea of communication. Chicago: University of Chicago.

Scheutz, Matthias (ed.) (2002): Computationalism: New Directions. Boston, Mass: MIT Press.

Sudnow, David (1972): Studies in Social Interaction. New York: The Free Press.

Turing, Alan (1936): On computable numbers, with an application to the Entscheidungsproblem. In: Proceedings of the London Mathematical Society, Vol 42 (1936-7).

Mobile communication and the public sphere

Moving and lingering: the mobile phone in public space

Joachim R. Höflich & Georg F. Kircher

People and media in motion and the privatization of public space

Mobile media allow people to be on the move and, simultaneously, to get in or to stay in touch with others. This progress has been advancing swiftly. People who move in public while being on the phone have become a common sight. Findings from observational studies suggest that people who use mobile phones have found ways to adjust themselves with respect to their behaviour in public settings and towards others (cf. i.e. Höflich 2006; Lasén 2006; Okabe / Ito 2002; Fortunati 2003; Murtagh 2002). Drawing upon these studies, a substantial study about the use of mobile phones in public settings was conducted. All following arguments are based upon this study.[1]

Behaviour in public settings is not a new topic of research. Georg Simmel, for instance, already examined the impacts of urban structures on mental life at the beginning of the 20th century. In his work *Die Großstädte und das Geistesleben* (Simmel 1903) / *The Metropolis and Mental Life* (Simmel 1950) he describes the basic attitude of publicly acting people as distanced, even reserved: "This mental attitude of metropolitans towards one another we may designate, from a formal point of view, as reserve" (Simmel 1950: 415). More explicitly, he describes this behaviour at the following point:

> "Actually this indifference would be just as unnatural as the diffusion of indiscriminate mutual suggestion would be unbearable. From both these typical dangers of the metropolis, indifference and indiscriminate suggestibility, antipathy protects us. A latent antipathy and the preparatory stage of practical antagonism effect the distances and aversions without which this mode of life could not at all be led" (ibid.: 416).

Looking at behaviour in public settings, one has to include potentially present attendees, to which the agent adjusts. While in the private domain the realization of individual goals is on the foreground, in public settings harmonization is required. The individual participation in social life in public settings does not draw upon understandings; the basic pattern, in fact, is the contemporaneity or coexistence of people. Hence, individual self-fulfilment in public settings demands

1 This project can be found under the title: "Mobile communication, telematization of everyday life and the change in media practices". The project was conducted between 2006 and 2009 with the support of the German Research Foundation (DFG) at the University of Erfurt, Germany.

complex spatial and social arrangements which make it possible to experience communality and to keep social distance at the same time (Weiß 2002: 34). Hence, the individual has to apply a different standard to its actions and it can only take, on a general level, certain practices of its environment for granted. However, one sees oneself confronted with being perceived – and the perception of being perceived has got an ongoing communicative relevance: When a person perceives that he is being perceived and that his perception of being perceived is being perceived, he has to proceed on the assumption that his behaviour is interpreted as adjusted hereupon. Therefore Luhmann argues that in interaction-systems it is not possible that one cannot communicate. To avoid communication absence is the only way (Luhmann 1984: 561 et seq.). Thus, among communicators in public situations a clear and predominant anticipation is that one can hardly evade communication or interaction, even a certain involvement is virtually presumed (cf. Goffman 1966). As individuals are exposed to looks of people who are unknown to them they have to adapt themselves to the fact that their behaviour is constantly exposed to reactions of people whose response the individuals cannot predict exactly (Bahrdt 1969: 56). Simultaneously, people strive to avoid embarrassment that results from their actions and behaviour because normally only an informal and beneficial interaction is regarded as natural. Embarrassment in reverse represents deviation from normality (Goffman 1994: 106). The disturbances of a communication-situation can evoke such embarrassments and are conceived as unpleasant. Hence, in public settings people try to encounter other unknown people preferably superficial in order not to give rise for a disturbance of the communication-situation: Unlike in private settings, in public settings only minor personal information are disclosed. According to Goffman (1971) arrangements on the basis of "civil inattention" can be seen as a strategy to adjust one's own action to others and, simultaneously, as a pre-requisite for privacy. Thus, acting in public settings is based upon the demonstration of an active participation in life, but, simultaneously, one pretends not wanting to interfere with private matter of others. Virtually, we eavesdrop without dissembling this. Herein, one can clearly notice the reference to Simmel's (1950: 414 et seq.) description of "reserve" or "blasé attitude" in urban, public communication.

An individual moves in its environment in a way that it definitely perceives its environment, and it does not contemplate its environment as long as the environment appears to the individual to be "natural" or "normal" (cf. Goffman 1971: 238 et seq.). People who act in public settings are perfectly aware of their reciprocal perception, however, short interaction often occur en passant.[2] A

2 Rules of distances allow us to avoid uncalled interaction (cf. Häusermann / Siebel 2004: 57). Mentioning previously that one has to be always aware of interaction, thus the communicative distance is in no contradiction to it. "While at the immediate and superficial level, encounters on the street are hardly noticeable and devoid of pleasantry and warmth,

taken for granted is what makes everyday life. Only special occasions, which diverge from the expected normality, forces people to give up his / her individual occupation in favour of a situational clarification. "Matters that the actor has become accustomed to will receive a flick or shadow of concern, one that decays as soon as he obtains a microsecond of confirmation that everything is in order; should something really prove to be "up", prior activity can be dropped and full orientation mobilised, followed by coping behaviour" (Goffman 1971: 238 et seq.). This "dissociated vigilance" guarantees the mastering of public situations. If one had to pay one's full attention to each and every communication situation a regular public life would become impossible since the numerous communication and interaction incentives would deter from a focused action. This can only work when disturbing influences do not gain the upper hand and thus determine public acting. Such influences are communication situations in which one is compelled to far exceed one's proper extent of attention. A mobile phone call can be indeed one of these influences. Drawing hereupon, Murtagh (2002) observed the behaviour of telephoning-train travelers and noticed that in those situations people avoid direct eye contact and try to pay the least attention possible to each other.

> "Goffman identifies what he refers to as 'civil inattention' as one of the rules that govern our behaviour in public places. [...] In public settings, Goffman argues, a state of mutual gaze is avoided. And so it was with phone use – mutual gaze between phone user and co-present was generally avoided" (Murtagh 2002: 84 et seq.).

The one on the phone can well be regarded as a disturbance – an annoyance he / she does not intend to be. Besides averting one's gaze, may it be out of embarrassment or may it be to indicate a private situation, the temporal disturbance of the phone call (even the ringing of the telephone) gives rise to normalization-testimonies in the form of non-verbal gestures which suggest that the disturbances are of a short duration and came unplanned, and thus, normality will be restored shortly (cf. Höflich 2009).

Media usage in public settings has become a topic of research (cf. i.e. Krotz 2007; Burkart 2000, 2007; Höflich 2005a; Höflich / Hartmann 2006; Hartmann / Rössler / Höflich 2008; Katz 2006, 2009; Ling / Campbell 2009; Hampton / Livio / Sessions 2009). Especially, the question of how mobile communication is integrated in everyday life and thus in public situations and whether or not this leads to disturbances or to (new) arrangements was discussed in these studies. Acting in public settings seems to be an ongoing process of social negotiation which, simultaneously, induces change. It remains to be seen, for instance, whether the strategy of "civil inattention" still works in its fundaments with the spread of mobile communication and whether or not previous rules still apply or

---

pedestrians do, in fact, communicate and do take into account the qualities and predicaments of others regulating their behaviour" (Wolff 1973: 48).

new ones are necessary. Does the usage of a medium, which is used for private and individual communication, contradict the structure of public spaces? Is the coexistence disturbed by the conspicuous, partially even extensive media usage? At least, when considering empirical findings, one can assume that people who use media while moving in public settings have a certain sense for the setting and its formal and social structures, and thus act accordingly (cf. Höflich 2005: 160 et seq.). Furthermore Löw emphasizes the reciprocity between action and setting, especially when she quotes that settings are not only experienced physically but settings also re-act on bodies. And thus, they structure action in the form of institutions (Löw 2007: 62). With reference to (mobile) media one can continue and say: "On the one hand, their [media] use is shaped by influences in that public space. These range from the spatial structure to the presence of others. On the other hand, the use of the media also has an effect on the public space" (Höflich 2006: 20).

Privacy in public settings is a subject which especially in relation to the mobile phone, is in the foreground. Simultaneously, privacy and public are basic principles of urban social life (Soziierung) (Bahrdt 1969: 83). Telephoning used to be a private and intimate domestic issue. With the introduction of mobile phones telephoning has become a public issue. Even if they are not only private matters which are carried into public, the mobile phone has contributed to the promotion of privatization of public settings right up to a tyranny of intimacy (cf. Sennett 1974 and i.e. Höflich 2005a, 2005b, 2006; Höflich / Hartmann 2007). This emphasizes once again that a spatial definition of privacy versus public is insufficient. Private matters are always carried into public, or respectively in this case carried around, it is, virtually, like a layer that surrounds us – "a thin layer of private space around the bodies of the people with whom we are sharing non-private space", as Lofland (1989: 12) denotes. This layer is more or less always permeable. Everybody who acts in public settings discloses something of oneself. Intentionally or not it cannot be avoided in everyday life. Especially while acting in public settings there is information which is meant for others and therefore meant for being disclosed, and there is some other information which should remain private and is not meant for disclosure (Weil 2005: 112). According to the "civil inattention" of Goffman, not everybody is interested in such private moments. There is something like a "conversational preserve":

> "The right of an individual to exert some control over who can summon him into talk and when he can be summoned; and the right of a set of individuals once engaged in talk to have their circle protected from entrance and overhearing by others" (Goffman 1971: 40).

Almost unavoidably, others quickly become an audience, when they did not want to be in the first place, if we disregard, in this context, voyeurs who take an enormous pleasure out of having an insight into things that are actually none of their business. Present third parties are each included differently into a person's communicative intents. For once, there are those people this person is having a conversation with. In the case of mobile phone use we deal with a communicative opposite number who is physically not present. In return, present third parties (which also apply for non-media contexts) are each differently anticipated: Not only it is taken notice of the auditor as an overhearer but he is deliberately involved into communicative proceedings. However, it is taken notice of the overhearer, but he / she is not voluntarily taken into consideration. An eavesdropper is like a voyeur neither considered in nor been taken notice of. This one eavesdrops anonymously in the background. This suggested distinction of people present by Bell (1984) can be categorized as; are the people known, ratified and addressed. This idea can be summarized in the following way:

Hierarchy of Qualities and Auditor Roles (Bell 1984: 160):

	known	ratified	addressed
Addressee	+	+	+
Auditor	+	+	-
Overhearer	+	-	-
Eavesdropper	-	-	-

Overhearers, especially those unknown ones that act in the background, deprive communicators of their control over the communication situation – and leave people vulnerable. Things only remain private as long as they do not leak unchecked through to others. To put it differently: access control is the condition to see something as private. So the protection of privacy means a protection against the uncalled access of others (Rössler 2001: 23). This aims at a definition of privacy as an inter-personal process of access boundary-control. In this very sense, for Irwin Altman (1975: 11) privacy represents "an interpersonal boundary-control process, which paces and regulates interaction with others". And thus, he defines privacy: "as selective control of access to the self or to one's group" (Altman 1975: 18). In other words, selective control is an integral part of such an understood privacy. People control how open or tight-lipped they are towards others. This refers to verbal, to non-verbal as well as to behaviour in a concrete environment and last but not least to cultural practices. In this, privacy has to be understood as a bi-directional and dynamic process; however, what is defined as private is time-dependent and can change depending on the respective circumstances (cf. also Altman / Chemers 1980: 77). Now, Altman combines privacy and spatial behaviour, or more concrete: he associates privacy with the possibility to keep personal space under control. In Altman's words (1975: 54):

> "...privacy regulation is perceived through a series of behavioral mechanisms, including verbal and para-verbal behaviors, nonverbal behaviors including the body, and environmental oriented behaviors of personal space and territory. This personal space is a mechanism used to regulate interpersonal interaction and to achieve a desired level of privacy."

Such a transformation from personal space into physical distances is closely linked to the name of Edward T. Hall and the term proxemics – or the study of how people use space as a vehicle for communication. According to Hall (1966: 1): "Proxemics is the term, I have coined for the interrelated observations and theories of man's use of space as a specialized elaboration of culture". Hall, however, does not want to move distance as such into the center of attention. He rather refers to the respective communicative possibilities which are linked to it. Body contact, olfactory signals and skins structures are relevant in situations within short distance, this changes when the distance increases. Hall distinguishes four zones of distance: intimate, personal, social and public distance.[3]

Especially in public space, one can observe people's effort of claiming territories for themselves or respectively the thwarting of such an intrusion. To this belongs Goffman's description of "stall":

> "The well-bounded space to which individuals can lay temporary claim, possession being on an all-or-none basis. A scare good will often be involved such as a comfortable chair, a table with a view, an empty cot, a telephone booth. In the main, stalls are fixed in the setting, although, for example, at beaches devices such as large towels and mats can be carried along with the claimant and unrolled when convenient, thus providing a portable stall" (Goffman1971: 32).

Such stalls are meant to visibly indicate that one claims a certain space as possession and one is determined to defend this spatial claim. It is expected by others, however, that a certain user-space is guaranteed. This refers to: "the territory immediately around or in front of an individual, his claim to which is respected because of apparent instrumental needs" (Goffman 1971: 34 et seq.).

---

3 The four zones are (again separated into a proximate and distanced zone) (cf. Hall 1966: 116):

1. Intimate distance (0-15 cm / 15-45 cm) – often accompanied by body-contact (i.e. sexual intercourse or wrestling) and not always regarded as appropriate.
2. Personal distance (45-75 cm / 75-120 cm) – generally, the usual distance between us and others.
3. Social distance (120-200 cm / 200-350 cm) – i.e. the distance of people who work together or of people who wait for a bus.
4. Public distance (350-700 cm / cm or more) – i.e. situations with a formal character, where also non-verbal signs are important.

This also includes, last but not least, the space which is regarded as necessary for an (undisturbed) telephone conversation.

Gender-specific aspects of personal space used to be a preferred topic of research. These studies especially refer to the fact that men claim much bigger personal spaces than women and that it is of relevance whether the opposite of women is a man or another woman; thus, women communicate with much shorter personal distance among each other compared to cases where men and women communicate (cf. i.e. Mühlen Achs 1993: 103 et seq., 2003: 193 et seq.; Henley 1977: 27 et seq.; Wood 2001: 146 et seq.). Besides, the personal space of women is more vulnerable than that of men: Thus, Henley records:

> "Despite the fact that women take up less space in many ways, there is evidence that what space they have is violated more than men's is. For example both females and males have been found to approach females more closely than they approach males" (Henley 1977: 39).

Furthermore it has been explored that women who walk on the pavement rather dodge men than vice versa. Generally, women seem to have a higher sensibility towards non-verbal signs (Henley 1977: 42). However, the findings are again not that clear. If one, according to Altman (1975: 75), takes into consideration other characteristics like age, ethnic origin, and situational factors then suddenly "the whole sex picture clouds".[4] In situations that are perceived as negative, for instance, people tend to take up more space than in positive ones (cf. Schulz-Gambard 1996: 328). However, this does not mean one can abandon the gender-aspect altogether.

As mentioned previously, research has already occupied itself with spatial respectively spatial-related aspects of mobile phone usage, starting from the use of niches as virtual phone booths, to the occupation of space as a temporary claim of private possession till non-verbal hints (giving looks, body language) during a phone call. These studies are more impressionistic than systematic. This holds especially for distance behaviour. A small and rather explorative study was conducted by Rich Ling (2002: 61, 2004: 134 et seq.). He speaks of a "good-natured experiment". This study deals with, as Ling calls it, the colonization of the public sphere by users of mobile phones and especially how people create their own territory in order to have an undisturbed telephone conversation. In an 'experiment' he intruded upon the personal sphere of callers. In another one he purposefully tried to make eye-contact with a person who was

4 Thus, also Schulz-Gambard (1996: 331) remarks that lots of findings of research about personal space seem rather inconsistent and weaker, than it is reported about in the (often very selective) secondary literature.

talking on the phone. The result was that these people tried to protect their personal space by sidestepping in order to avoid the intrusion.[5] A conclusion:

> "Unlike the metaphor of the boorish ego maniac loudly talking on his (sic!) phone, this experiment seems to point that mobile telephones users are aware of the need to maintain the space around them if possible" (Ling 2002: 76).

These people felt that a distance rule was violated – and without explicitly sanctioning the intruder's behaviour they got out of the way with the intention of restoring the 'normal' personal space zone. However, Ling does not go into dwell on gender-specific aspects. Although, the topic sex or respectively gender has often been a topic of research, starting from gender-specific use up to mobile phones as accessories, however, studies about the distance behaviour of mobile phone usage with regard to gender-specific aspects are missing.

Scenes of event

As a starting point for the exploration of public communication situations, we defined what constitutes (urban) public settings in order to receive criteria for the selection of observation scenes. Schubert (2000) offers helpful classification as an orientation in his comprehensive classification of public settings. Of the twelve different "settings" which differ in relation to public towards privacy as well as in the degree of structuring, but only four of these settings are particularly relevant for this field of study: *Routes of Transport and their Edges* form dynamic elements in urban structures. Especially, bus and tram stops are places of lingering and transition. *Mobile Rooms of Transit* like elevators, busses, and trains are more or less freely accessible, but represent in themselves a closed, semi-public setting which again uses public routes of transport. A special focus lies on *Local Urban Centers*. These are places one would typical regard as public settings. Limitations on access and earmarking are relatively low here. Classic examples for these kinds of public settings are big open squares or pedestrian areas in the city center. Closely linked to the previous are *Surroundings of Consumption Places*. Those public settings are mainly characterized by commerce. Thus, a more clearly earmarked situation arises than in these local urban centers (cf. Schubert 2000: 55 et seq.). Burkart (2000) again classifies public situations with the criterion of how communication is arranged. Here, Burkart classifies into three categories which can also be laid as a raster over Schubert's classification of settings: Heavily regulated situations with

5 This is not unlike those reactions people show when they intrude on each other's personal distance zone. However, not only people on whom others intrude feel uncomfortable with such an intrusion. The same holds for the intruders (cf. Altman 1975: 90).

specific functions, "mobility gates"[6] in which communication is regulated to different degrees and situations which are rather weakly regulated by unspecific functions or multi-functionality. For Burkart, weaker regulations of communication are accompanied by diminishing disturbances of mobile phones (Burkart 2000: 222 et seq.). Besides the mentioned social functions of settings, Burkart also includes the spatial features of place as well as its communicative and social rules. Thus, the size of a place and the acceptance of certain communication forms are a crucial factor to whether or not mobile communication is disturbing (ibid.).

Multi-method procedure

Observation and survey

A methodical approach on a multi-level scale was chosen to answer the question of how mobile communication is incorporated into public cohabitation. Firstly, public situations of communication have been observed of over a longer period of time. Places, patterns and behaviours could thus be observed from an external perspective. Within the course of two observation stages 370 observation units were recorded, 270 of which were in 2006 and 100 in 2007. While the thematic perspective remained the same, the methodical procedure was extended: While in 2006 an observation unit was made up of an observation protocol, in 2007 the observation were conducted simultaneously by two people and subsequently, the people observed were briefly interviewed. Thus, such an observation unit is made up of two observation protocols and one interview protocol. Altogether, we had 570 single documents for data evaluation. The conception of these observation protocols reflected the methodical perspective of this study. The observation sheets were designed openly, besides, the thematic focus on mobile communication and the standardized fixation of formal data (i.e. time, weather conditions, and place). In 2007 we had the additional option to specify selected subjects by conducting subsequent interviews and thus add evaluation of the subjects. Additional information was obtained by the addition of these double observations. The written recordings were completed in every stage by video and photo recordings. Such visual recordings which can also be found in the observation studies of Humphreys (2006) or Höflich (2006) serve especially as an illustration of findings or as it is the case in the long tradition of visual anthropology (cf. especially Collier / Collier 2004) as an own source of data. Visual sources are the result as well as the starting point of the empirical work:

6 The term "mobility gates" is used as a translation of the original term "Mobilitätsschleusen".

Certain video and picture sequences, which become apparent time and again later on, are like the proverbial golden thread which can be pursued through all stages of the survey and which also connect the respective methods with one another.

Group discussions

In order to extend the observation data with the perspective of the people acting we staged group discussions as our subsequent methodical step. This subjective insight seemed necessary not to substitute the social reality by a reality which a scientist had constructed (Schütz / Parsons 1977: 65 et seq.). Dreher claims discussion to be the exclusive method to understand individual perspectives in the perception of reality (Dreher 1995: 187) and thus this was the subsequent expansion of the more objective external perception of observations and interview situations during brief surveys. The criterion herein is not the exclusivity of a perspective but the complementation of both. For the group discussion we selected participants from different age-groups in order to collect opinions from people from all walks and stages of life. The youngest participants were school students between 16 and 18 years of age and the most senior participants were up to 65 years of age. Altogether six group discussions with a number of five to six people were conducted. In these discussions the participants had the chance to explain their perspective on their own acting as well as on the acting of others. Against the background of the survey data, the observation data could be re-discussed. One group discussion took 90 minutes and it was recorded, transcribed and analysed for evaluation purposes. For every group discussion a written record was kept in addition to the sound recordings, in order to highlight features that were imperceptible on the sound recordings. The group discussions were structured on the basis of a guideline which covered different issues of daily routines, public and mobile communication. We, however, made sure that the researcher interfered only discreetly in order to not interfere with the discussion-dynamics. As an opener for the group discussion and as an introduction for the subject we asked all participants to present their mobile phones and to talk about themselves and their relationship to mobile media. For the purpose of a later evaluation we photographed the mobile phone and the use of it. While it was relatively easy to record spatial behaviour in the stage of observation, we met some challenges in the stage of group discussion. Thus, a discussion basis had to be laid on upon which mutual observation to spatial behaviour could be exchanged. In doing so, a notion of behaviour in space – a thing the participants generally perform subconsciously – should be evoked and made conscious. For this purpose we embedded the picture and video-sequences from the observation stage into the group discussion and used them as discussion stimuli. The participants now could comprehend the previously conceptionally derivated arguments on the basis of concrete examples and which they then could again discuss. Thus, the dynamics of a discussion can evoke vivid statements by mutual stimulation (Dreher 1995: 186). However, in this case the "stimulation" did not come exclusively by the other discussion participants but also by the visual incentives.

Survey

A lot of questions regarding the relation of private and public moments of mobile phone use were taken up within the framework of the group discussion. Drawing upon the close relationship between the uses of the mobile phone, the perception of privacy was explored by a paper-and-pencil interview. Altogether 177 people, 100 male and 77 female (average age 38), were surveyed on the basis of a quota target. The survey's underlying assumption was that a possible shift between private and public use of mobile phones cannot be seen as detached from a comprehensive medial development. Simultaneously, questions to the use of mobile phones in public settings were in the foreground in our survey. One of these aspects is personal space behaviour which is especially relevant in connection with the claim of private spheres in public settings. Especially the later following findings draw hereupon. Approaching the subject of personal space behaviour within the frame of a paper-and-pencil interview is methodically not unproblematic. Especially in connection with the observation, space and distance behaviour get under close scrutiny. Again, with the group discussion one could enlarge upon these moments. However, some need for clarification remained, especially, in connection with gender-specific communication patterns. A paper-and-pencil questionnaire was meant to shine light onto these matters. For that we pursued two methodical approaches:

1. It was drawn upon the 'Comfortable Interpersonal Distance Scale' (CID) by Duke and Nowicki (1971) in order to measure interpersonal distance. The intention of the scale is "to provide a psychometrically sound measuring device within an acceptable, standardized, manipulable, paradigmatic methodology, for the study of interpersonal distance and its parameters" (Duke / Nowicki 1972: 119). Technically speaking, this is a placement scheme without a scaling target. The assessed distance had to be marked as a circle on a line whereas the distances are measured with a ruler in millimeter-terms. The underlying scale for our survey has been slightly modified and looks the following way:

Please, imagine you are having a personal conversation on your mobile phone in public. What is the minimum distance to other people that would still make you feel comfortable?
Please read through all people listed in the left-hand column and then cross the respective lines in the right-hand column to indicate how large the distance to them should be. The point at the beginning of the line is you.

My personal distance to…	
… an unknown person (female)	●————————————
… an unknown person (male)	●————————————
… a distant aquaintant (female)	●————————————
… a distant aquaintant (male)	●————————————
… a female friend	●————————————
… a male friend	●————————————

Two dimensions were explicitly incorporated herein, for one, the kind of relationship to those people who surround one during a telephone call, and for another we varied between the facts whether the other people are male or female. The proceeding we deal with here is a projective test, that is, the people questioned had to assess an abstract situation and this assessment was asked to be sketched, in their personal conception, on a sheet of paper. However, projective tests are under a special reserve because their reliability is not very high and their correlation with non-projective test is rather low (summarizing: Roeder 2003: 52). Non-projective tests are again based on the assessment of real and / or real-spatial distances. Within the frame of a survey, such a test is used with the aid of a picture test.

2. The people surveyed were shown four photographs each showing different distances between the person on the phone and a present third-party in order to assess the different distances within the frame of mobile phone use. The photographs were chosen in reference to real observation situations. However, these photographs were not taken during the observations. In fact, these photographs were staged in order to maintain the anonymity of the people observed. The pictures were refined in a multi-stage procedure: First, the distances were intuitive reconstructed. Subsequently, the pictures had to undergo a pre-test. It proved that exactly those photographs were chosen which correspond to Hall's interpersonal distances (see footnote 3). Besides, the photographs were used during the group discussions as a discussion basis. Subsequently, after these two test stages the photographs were incorporated into the questionnaire the following way:

In everyday life we often have close encounters with people we do not know personally. In the following four pictures you can see a person (A) who is window-shopping and another one (B) who is talking into his mobile phone. From picture to picture the distance between them diminishes.

We now would like to know, from which distance on you personally would feel uncomfortable. Please, give the number of the picture which fits best to the questions below.

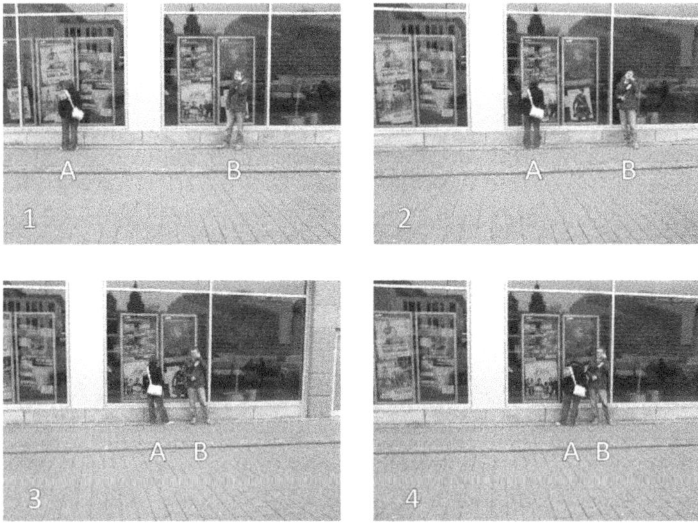

Please, imagine you were in the situation of person A. From which distance onwards you would feel uncomfortable?	1☐ 2☐ 3☐ 4☐ none☐
Please, imagine person B would be female in this situation pictured: From which distance onwards you would feel uncomfortable?	1☐ 2☐ 3☐ 4☐ none☐
Please, imagine now you are person B who is talking on the phone. From which distance onwards would you feel uncomfortable in this situation?	1☐ 2☐ 3☐ 4☐ none☐
Please, imagine now person A would be male in this situation pictured: From which distance onwards would you feel uncomfortable in this situation?	1☐ 2☐ 3☐ 4☐ none☐

Contrary, to the assessment within the frame of the (open) scale the people questioned were asked to project themselves into different situations: First into the situation of a present third-party and then into the situation of the person talking on the phone, additionally it was anticipated that the third-party present was either male or female. Despite all scepticism about projective tests, it proved that these two methodical steps complemented one another usefully, especially because one method built upon the other and thus, evaluation coherence was given within the frame of the questionnaire.

Description of observation scenes

As observation settings were such settings chosen which reflect public life and where we could draw upon the assumption that people would use their mobile phone there. The settings are based on the classifications of Schubert and Burkart. Thus, a cross-section through settings within the German city of Erfurt was constructed. Especially, one setting got into the focus of our observation interest: The Anger square in the center of Erfurt. Since medieval times, this square was and still is the central place in the center of Erfurt. A large proportion, or nearly a third of all observations, was made at the Anger square which is owed to the fact that this setting fuses to an especially high degree different "sub-settings". However, the following holds true once again: Space is a basis and correlate of human behaviour which is constituted of motions into certain directions and certain distances to objects (Kruse 1996: 314). This big and mostly busy square is separated by tram tracks. The tram stops are again settings of lingering and waiting in the center of the square. Additionally, there are several areas with public benches as well as chairs and tables from bordering street cafés. Characteristically for the motions on the Anger square is the large number of entrances and exits for pedestrians and trams. The eastern corner of the Anger is characterized by a large shopping mall which is again another access to as well as exit from the square.

The railway station, which is symbolic for places of transit, can be structured into three sections which again intertwine. The platform section is defined by waiting situations and the transition into the means of transport. People who disembark trains, again proceed via the platforms to the station's exist. This sort of motion continues throughout the other sections of the station. These people who are proceeding for the exits encounter in the shopping concourse section, people who are passing their waiting time or who come to the station to do their shopping. By passing the station's atrium one reaches the station's forecourt where the previously often compact crowd diffuses. The tram stops within the city center as additional "mobility gates" (Burkart 2000: 221) are often so crowded during peak times that embarking and disembarking not always proceed frictionless.

The shopping-center Anger 1 which is located at the top of Erfurt's Anger square directly in the heart of the city, can be regarded as an example for a strongly commercial semi-public space. One of Erfurt's arterial thoroughfares goes past behind this shopping center. Thus, the building shields the Anger's pedestrian area from road traffic and with its underground car park it simultaneously functions as a gate between public space and road traffic. Especially, during peak business hours the main entrance is a bottleneck, since the people in the relatively broad area in front of the building have to channel towards the shopping center as people or groups of people often just stand or linger in front of the entrance area or, again, other people desert the building.

A large number of cafés are located around the nearby located smaller squares: Fischmarkt and Wenigemarkt. A lot of these cafés have outside seating. A special situation arises from the cafés SiJu and FAM which are located vis-à-vis. Their outside seating areas confine the already small pedestrian area so that at certain times of the day a thick and slow flow of people accrues. Furthermore, people time and again stop by the edges of the outside seating areas in order to converse with people seated in one of these two street cafés. Some people stroll in this so created alley between the two cafés while people who sit at the tables outside virtually present themselves. Especially, at night times an atmosphere of seeing and being seen emerges.

During the course of the study it more and more occurred that mobile communication could also proceed, in some cases, from public roadside phones. Especially, at the Anger such cases could be observed. Such phones are located in different areas of the Anger. For this reason we included such places into the observation study.

Results of the observation studies

The use of mobile phones could be observed in all the public places mentioned. Most of the people, who were moving while using their mobile phones, could be found on squares, or in rooms of transit, such as station concourses. Especially in these places, a high rate of transit traffic could be observed at certain periods of time. A lot of people just crossed these places (considering the fact that pedestrians tend to use the shortest way possible (cf. Whyte 1988: 57)), which created a certain dynamic. It was noticeable that people, who were using media in motion, tended to look at their mobile phones, or during calls on the ground, or they focused nearby objects. Such body posture and language suggests a high level of concentration on the medial communication situation. In this context, one may follow Gergens (2002) line of argument and say that the perception of the physical presence is undermined by the presence of the communicational situation. He argues that mobile phone users have to give up on large parts of the present situation, in order to be able to properly participate in the virtual "absent presence" which is created by the telephone call. This is also proven by the fact that such people walked slower than others. A focus on objects and their surroundings in public settings, as was also observed by Höflich (2006) on Italian *piazzas*, can hereby be understood as a simple orientation guide in the present situation. One may suggest the assumption, that movement during media use in public settings might be regarded as a disturbance, mainly just because navigation as a vehicular unit (Goffman 1971: 5) is only possible in a restricted way. However, the observation results point in another direction. No obvious conflicts between users in motion and the people around them could be observed in any of the observation stages. Follow-up group discussions will show whether or not these observations will correspond with the assessments of other people. Mobile communication does not seem to stand out from other actions on public places – if one follows Burkart's (2000, 2007) classification of settings and their influence on mobile communication – as these are marked by their "unspecific function" or their "multi functionality". It neither seems to have a disruptive effect on the "mobility gates", which are characterized by a higher level of regimentation according to Burkart's classification of settings. In spite of being embedded in a medial "absent presence" there seems to be enough attention left for the coordination of movement, which points to the continuing presence of a "sense of place" (cf. Höflich 2005, 2006, 2007). People are in any case aware of their environment on a basic level, so that even highly-embedded users showed impulses of stepping aside in order to avoid collisions or dangerous situations with trams for example. However, one would not do the observations justice, if they assumed that only media users possess feeling for the present situation, as a smooth course of action seems to be based on a certain amount of coordination on both sides. Within the framework of a mutual arrangement, also other people may therefore be aware of the "sense of place" of mobile communicating people.

Spatial orientation

Spaces are spaces of perception and action and function as starting points for targeted action and motion (Kruse 1996: 315). Out of the observed public settings, the Anger square in Erfurt offers the widest range of opportunities for usage and action. It can be divided into numerous smaller settings which have different characteristics respectively. This also affects how users of mobile media move in public settings. In most cases observed, there was a consistent kind of movement during media use. However, there were also cases which were characterized by a change between two or more different status of movement. For instance, a kind of "stop and go" could be observed on a woman while she was using her mobile phone:

> "She takes out her mobile, lowers her gaze and looks at her mobile; stands still for a short moment and looks at her display with her head deeply bowed. She types on her mobile, continues walking slowly and pauses for a few more times on her way to the entrance. About ten meters in front of the entrance she looks up again, raises her head, shuts her mobile and walks with increasing speed into the train station, her mobile in hand".[7]

Because of the fact, that it was not possible to observe the begin of an action in each of the situations observed, it is hard to tell whether or not mobile media use

---

7 Quoted from the observation protocol. Every following quotation in this paragraph will also be from this protocol, without explicitly referring to it again.

in public settings may or may not be characterized by a change in movement patterns, or if the observed tendency of keeping a certain kind of movement pattern, which has been deemed as acceptable, is correct. This would be supported by the findings of Whyte (1988: 66): "What one is walking past influences how fast one walks." However, a change in movement patterns after mobile media use could be observed in many cases. Thus, a lot of people start driving or walking after they have finished their call and put away their mobiles, mostly into the pockets of their jackets or trousers. One male cyclist for example "[...] finishes his call now, looks around, puts his mobile into the front pocket of his jacket and mounts his bicycle". Other people again pause or sit down when they were walking beforehand. It could also be observed that some people only entered a shop after they had finished their call, which will be described later on. Mobile phone users can also adopt a different movement pattern than their companions, when switching between two movement situations, as described in the following case: A man and a child were observed which were both walking towards a café. "When arriving at the café the child enters the building while the man stays behind in front of it. After a short time he finishes his call and also enters the café". Another special way on how to orientate yourself on a lively square even though you are also concentrating on your mobile phone, is being shown in the observations: Hereby the observed people benefited from their companions who were walking in front of or next to them, and who were used as guiding points in order to be able to find their own way through the crowds – this showed that a vehicular unit can always also be a participation unit (Goffman 1971: 19). The own necessary ability to navigate yourself through the crowd has in this case been outsourced, so to speak:

> "Her friend is walking a bit faster and guides her somewhat. She is looking straight ahead. As she reaches a street, she wants to continue walking, but her friend stops her".

By relying on her companion, the medially embedded person is able to concentrate harder on the medium.

Aside from people who cross the square and are defined by their purposeful walk as well as by their spatial orientation, another group can often be observed. In almost all of the observed spaces there were people who were moving, but did not seem to have a certain destination. They stroll while using their mobile phones, walk without any evident purpose around objects or go in circles or spin around. Although in many cases it could be found out a posteriori that these persons had a certain starting point and were heading for a certain direction, it could often be observed that during media use they did not use the shortest way in order to reach their destination. The movement pattern of one man who was taking small detours on the way to his destination is exemplary for that:

> "He phones and walks from the direction of B3 [Bench 3; the author] towards the post office / tobacco shop within the post office building. He first walks at an average speed

> [...] but slows down considerably when approaching B2. Instead of using the shortest way, he takes a detour around the benches."

It seemed like he wanted to finish his call first, before entering a new situation. While walking and navigating around the square was possible during mobile media use, it seemed as if a change in place in the described situation also provoked a change in use: The newly entered room demands another set of rules. As it is a new behavior setting (Barker 1968), it apparently demands new behavior patterns and is not used until these new behavior patterns have been adopted. And with the same attitude that takes calling and typing for granted, mobile phones are tucked away when entering such rooms, so that something similar to an *etiquette of the street* (cf. also: Anderson 1990: 210) can be assumed.

Because of the fact that these non-focused users use their mobile phones for making calls or writing text messages on the square in such a natural way a certain habituation has to be assumed. They stroll for everyone visible; whereby objects on the square, such as flower pots or benches, are used as orientation points (cf. also Höflich 2006). However, also "virtual" reference points, such as the center of a small circle around which the person is moving, were often observed. It seems plausible, that these orientation points do not call for a higher level of concentration on other people while orienting yourself, as the circling of such objects often takes place in less frequented areas of the square. Passers-by may also interpret this behaviour for their own navigation, by noticing the signals of mobile media users and thus navigating in such a way as to avoid a confrontation. Our observations therefore do not show, except for one case, any cases of passers-by that have been disturbed in their motions by mobile phone users. Due to the fact that mobile phone users often move slower and are looking down to the ground it seems necessary to move around them. The fact that they move in a public setting seems to be less relevant than the actual phone use, which again is an indication of Gergen's concept of *absent presence*. However, there does not seem to be a conflict of regularly systems between mobile media users and passers-by, as Burkart (2000: 229 et seq.) assumed there would be in certain public situations and constellations. How strongly the desire to not disturb other people can clash with being absorbed by mobile media use, shows the observation of a man who crosses a square while typing on his mobile phone:

> "When two passers-by closely walk past him, he raises his head and tries to get out of their way although this would not have been necessary. By doing so he does actually get in the way of the passers-by who really need to step aside now. After that he looks at his mobile again and continues walking in small steps while typing."

Being suddenly interrupted from his absorbed mobile phone use, he "overcorrects" his behaviour. When concentrating on media use, space is only registered as a structuring element and environment blanked out completely except for the most basic perception of it – a behaviour which can only be interpreted by taking mobile media use into consideration. Without the recognizable use of media, such movement in public settings would seem strange to outsiders.

The use of mobile phones does usually not seem to be much of a problem when crossing squares. It was often observed that people, who seemed to be rather absorbed into their actions, looked up or around them every once in a while in order to orientate themselves on the square. Where this could be observed, it can be said that these users often walked in other people's wake, or – if the situation permitted it – walked across free spaces where they could not get into anybody's way. Such behaviour patterns were described by Goffman as early as the 1960es and 70es and can also be observed in connection with media use during movement. It can only be assumed, in how far these traffic-flows work because of the increased attention of the user, who has to master two situations at once, or because of other passers by who consciously or non-consciously pay to a certain degree attention to these users and therefore have a part in making the flow work (cf. Goffman 1966). The following is a good example for the arrangement with other people on a square:

> "She comes from the direction of the *Anger* square; carries a bag by the label *Tayler* in her left hand, wears a *Puma* bag on her left shoulder and uses her mobile phone with her right hand. She stops briefly in front of the train tracks when a car drives by and continues talking on her phone. She looks around a bit and continues walking straight ahead past the shopping mall *Anger 1*. She has to dodge some people who walk past here, but walks relatively fast and purposeful. She goes around the corner at Anger 1, crosses the train tracks again and goes into the direction of a shop called *Quelle*. [...] She continues walking, still talking on her phone. She has to go around a larger group of people and walks very close past a car, in order to avoid them."

Perseverance

People who are standing while using their mobile phones often have a point of orientation. They stand in front of shop windows, next to park benches, pillars, signs, flower pots or lean against these objects. People, who were really standing freely in a public setting while using their mobile phone, could not be observed. These orientation points seem to deliver a certain degree of protection, therefore making the situation safe. However, unlike in earlier studies on mobile communication in public settings (cf. i.e. Höflich 2006) it could only rarely be observed that people actively look for niches to retreat to while using their mobile media, as was the case with a woman who stood behind a large plant covering her.[8] Hereby the fact whether or not a square has any places to retreat to at all has to be taken into consideration. Furthermore, can these big flower pots which people walked around not be regarded as niches per se, but rather as objects which are "claimed" by users as a kind of private territory for a certain amount of time. Apart from the previously mentioned places, people who write texts or phone while standing can often be observed at bus stops or on platforms. It seems as if periods of waiting can be bridged by using media. Oftentimes there are only short periods of usage which can be observed.

Overall it can be said that standing and strolling users are not very focused on their environment, but tend to look on the ground during telephone calls or on their displays. Because they do not move, except for maybe bobbing up and down or revolving on the spot, they are not in the center of public attention. There were no cases were users were regarded as a disruption to the movements of other people. However, it can be assumed that they might be an obstacle to people when they are standing in rooms of transit, such as bus stops and platforms, and are too absorbed by their activities or do not stop them in time.

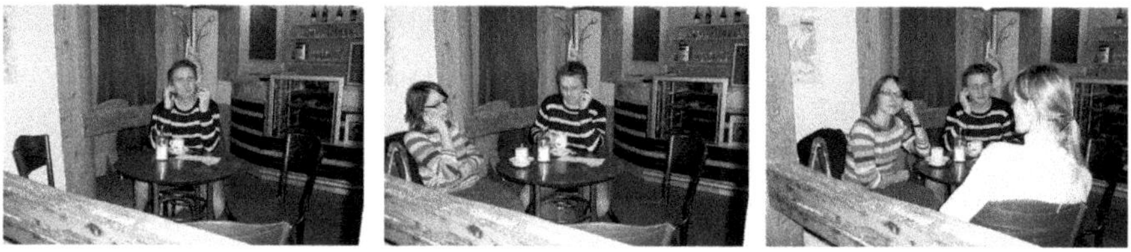

Re-enacted observation scenes in a café

Certain groups of motion were expected and could also be observed in cafés. During most observations people sat at the table when using their mobile phones; however there were also cases of people walking through the café. Parallels to the group of standing users can be assumed due to the lack of spatial movement. From time to time it could be observed that people were sitting in a café using their mobile media, and only had the sparsest contact with their environment. This is especially the case when people sit at their table all by themselves. However, this behaviour could also be observed when more than one person sat at the table. A woman who was accompanied by another person for example was strongly embedded into her phone call:

> "She phones with her mobile in her right hand [...]. Across her sits an elderly gentleman dressed in a blue shirt and light trousers. During her call the observed woman assumed a leaned back position in her chair. It is of notice, that her body does almost not move, but

8 Observation studies on Gender and Mobile Phones which have been conducted in Italy in May 2008 furthermore indicate that women tend to look for such niches more often than men. Results are not as clear though.

she incessantly plays with her left hand with her hair and wraps a long curl around her index finger. During the telephone call a waiter puts down a beer in front of her, without her showing any kind of reaction. Shortly before finishing her conversation she puts her hands on the table and drums the table. She then finishes the call and leans over the table. After that, she starts a conversation with the man sitting across her".

This movement of leaning over simultaneously reflects the movement from the virtual room of conversation to the physical and social room of the here and now.[9] However, it is more often the case that mobile media users take notice of their accompanying person during their mobile phone use and that they are even integrated into the call, by participating in the conversation via the user or by sharing mobile phone contents. Such situations are often characterized by a change between short mobile phone use and a turn towards the other person. Despite mobile phones usage there is an apparent connection to the other people sitting at the table. In such cases the motions of the observed people can be described as dynamic, even though they are sitting and therefore not moving in the room. In some cases mobile phones and their use become a communicative reference themselves and are the starting point of certain dynamics as in the following case:

> "After that she shows the mobile phone to the man and points to the display. He takes the phone back, presses some keys, puts it on the table after a while and leans back into his chair. When the man and the woman show the mobile to each other, they talk a bit to each other."

A place that structured the motions of the user on entering the room was telephone cells. They allowed people to make calls from outside their homes before the invention of the mobile phone (cf. Behme et al. 1988). Our observations enabled us to make out two basic patterns of usage: The first one shows an interruption of movement due to the usage of mobile media, as being shown in the following case:

> "The observed young man walks quite fast and determined from the direction of the tram stop to a public phone which is on the left-hand side of the observer. He hastily pulls his mobile out of his trouser pocket, puts some coins into the telephone slot and lifts the receiver. [...] The call is very short; the young man replaces the receiver and walks quite fast into the direction of the Anger 1 shopping center."

In such cases the use itself was rather short and hectic. Most users explained during the interviews that they have used a telephone booth, because their pay-as-you-go phone was out of credit. Their movement had to be interrupted for making a call on the telephone booth when otherwise they would have made the call from their mobiles while walking simultaneously. The second case that could be observed was that the use of telephone booths was part of another

9 cf. Article of Schlote and Linke in this volume.

activity. Especially people who already were in the proximity of telephone booths – mostly sitting on benches with a friend – were fairly relaxed during their call. They walked towards the telephone cells in a more leisurely way and their calls were longer.

Results of the group discussions

The next stage of the survey consisted of group discussions which allowed us to find out more about the perspectives of the acting people. As already mentioned, pictures and videos of the observation stage have been used as reference for the discussions. The scenes showed people in different public communication situations. For more detailed studies on mobile communication during physical motion especially one video sequences was of importance. Herein a young man uses his mobile phone while moving in a public setting. Thereby not only his mobile phone use (making calls, writing text messages), but also his movement patterns vary (strolling, circling, crossing). Although the video sequence was only staged, the scene had also been observed as such. During and after the video the participants had the chance to comment on what they had been shown. A lot of participants spontaneously said that they knew such a situation from their everyday life; some even recognized themselves in the media user. Others expressed that they were at first surprised by the video content, as they no longer registered such behaviour in everyday life. The participants as a whole nevertheless agreed on the fact, that they have been in such a situation themselves or have observed it. However, the scene, and hereby especially the actions of the young man, was quite differently evaluated by the single participants.

Disruption of public procedures

While watching the video some participants of the group discussion emphasized that they feel annoyed by people who move in such a way in public settings. Two elderly participants went as far as accusing the man of abnormal behaviour. "This is a person who doesn't recognize anything else and only lives with his mobile phone" (Quincy, 65). Another woman even said that she feels sad when she sees young people behaving like this. However, while most participants did not comment on the intensity of their feelings towards such behaviour, there were some participants who assessed it to be negative at first. Exemplary for this is the statement of a younger female participant: "Writing texts and probably getting in the way of other people at the same time. [...] I can't stand this" (Tina, 20).

The observation results at least proved the fact that most participants of the group discussion evaluated this scene to be everyday, natural and in this sense ordinary. It was often said that people recognize themselves in the behavior and actions of the young man. Exemplary for this is the comment of a young woman: "This could have been me in such a situation" (Katrin, 22). It was also added, that it often makes no difference whether or not people move in public settings with media.

Remarkably, two possible explanations for arrangements have been brought forward again and again by the participants in the different group discussions. Strolling around while making a call has been interpreted as a subconscious attempt of getting out of the way of other people. This is in alignment with the assumption that even people who are heavily embedded in medial communication still have a basic sense of their environment and the people around them. Another interpretation was that people move during media use in public settings, in order to create a kind of private sphere. Moving along objects or circling them was interpreted as signaling other people that they claim this territory. Ongoing movements simultaneously prevent too close of a contact with other people over a longer period of time. "To claim a territory, however, we must go through the rituals of establishing and agreeing upon illusory perimeters, or 'symbolic fences' (Goffman, Gullestad). We become quite accomplished at ignoring others who are in quite close proximity, through the use of fictive curtains" (Ling 2004: 125 et seq.). This primarily seems to work because other people can interpret this behaviour.

Reflecting own actions

It became apparent, that especially those group discussion participants, who made skeptical remarks regarding mobile communication even before they were shown the video; felt that movements in public settings were disruptive. Because of this, it was often difficult to establish whether these assessments resulted from their own experiences and values, or whether they resulted from present argumentation. For this reason we asked during the discussion whether the medium itself, or the movements during its usage, were the reason for these feelings of disruption. Those participants who spontaneously sensed a source of disruption in public settings in the cases shown, said after some reflection, that they probably would have acted in the same way in such situations. This change of perspective – relating to the people in the video – led to a relativization of opinion with some participants. Most of the participants, who said that the video showed a disruption of public movement patterns, increasingly tended to say, that they generally rarely felt annoyed by mobile communicating people in public areas. Instead of that, disruptions in higher regulated areas were assumed during the discussion. Against the background of these observation results these statements lead to the assumption that people, when asked directly about public behaviour, assume a disruption through the use of mobile media in some situations. Observations and inquiries during the group discussions however showed that mobile communication in public settings is oftentimes overshadowed by other actions and movements. The observations at least showed almost no conflicts between moving users of mobile media and passers-by.

Mobile communication and distance behaviour

There was no apparent connection between mobile phone use, distance behaviour and gender of user within the framework of the observations. However, the results of the inquiries were much clearer. We hereby used Goffman's (1971: 29 et seq.) definition of personal space: "The space surrounding an individual, anywhere within which an entering others causes the individual to feel encroached upon, leading him to show displeasure and sometimes to withdraw." When determining different distances, which simultaneously reflected Halls reaction zones, people were asked to keep in mind interactional dimensions by not only assuming the role of the caller, but also the role of present third-parties (as overhearer). Most interviewees said that, in regard to the last aspect, they felt uncomfortable when entering the personal space zone of the calling person. This is shown by the following illustration, which underlines the fact, that also intruders feel uncomfortable when violating the personal space of others: "Thus not only do people react to being intruded on themselves, but they are also sensitive to the personal-space boundaries of others; there is a strong feeling of discomfort when they are forced to intrude on others" (Altman 1975: 90). Furthermore it was shown that women feel uncomfortable much earlier than men (meaning as soon as the social distance zone).

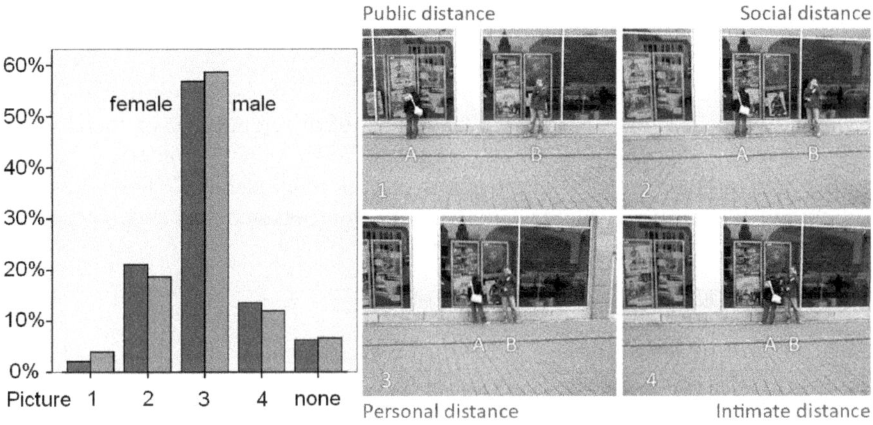

Illustration: Feeling of discomfort when intruding, in dependence on different distance zones during introduction (N=177).

This feeling of discomfort is much more obvious, when it is assumed, that the interviewees were having a telephone conversation themselves.

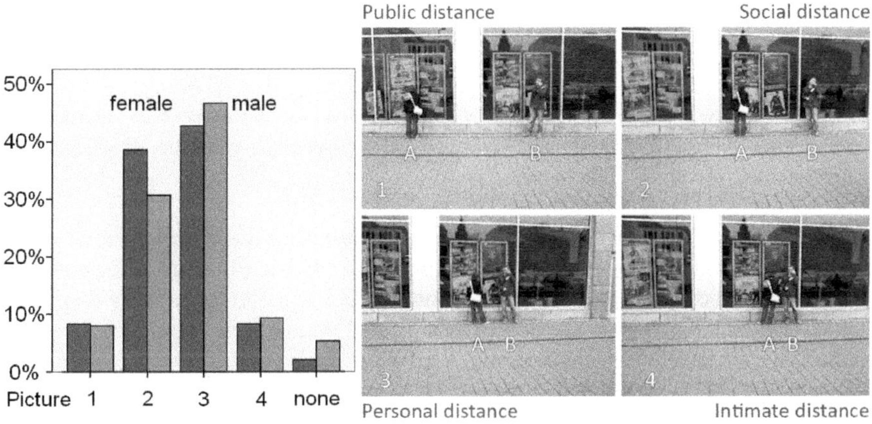

Illustration: Feeling of discomfort during a telephone call in dependency on different distance bubbles of an intruder (N=177).

Callers develop negative feelings towards an approaching person much earlier than intruders. This again especially holds true for women. A non-reciprocal relationship in the context of mobile phone usage can be manifested, whereby callers apparently feel a sense of discomfort sooner than intruders. This hypothesis is backed up by the assumption of non-reciprocal distance relationships in general, as for example stated by Henley (1977: 41): “A may not approach B to the same distance that B may approach A”.

Personal space depends on interpersonal relationships, according to Altman (1975: 79 et seq.): “Here we turn to the impact of different types of social bonds, As a mechanisms designed to help regulate self / other boundaries, personal space will probably be different across social relationships.” This should be examined with the aid of the Comfortable Interpersonal Distance Scale. The results are shown by the following illustration:

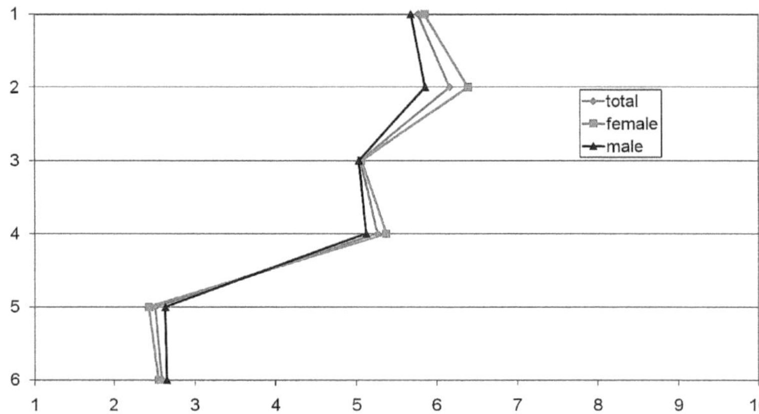

Illustration: Sense of discomfort during mobile phone calls in dependence on interpersonal relationships. (1=stranger female, 2= stranger male, 3=acquaintance female, 4=acquaintance male, 5=female friend, 6= male friend)

As might have been expected, the people surrounding the mobile phone caller during his call – unknown people, acquaintances, or friends – are of importance in regards to mobile phone use. Also gender-specific differences were displayed here: Women tend to start feeling uncomfortable sooner than men. This is not so much the case when the other person is a friend. Empirical findings which show that women keep a smaller distance to other people, especially to other women, within the framework of socio-emotional relationships (cf. Roeder 2003: 53) is not significantly shown, but there is a certain tendency pointing towards this. Mobile phones do altogether not have a huge influence on the definition of interpersonal distance zones. It rather is manifested, that mobile phones are not only embedded in the body of social rules of public communication but also confirm this fact.

At the end, we also want to look into the question which kinds of topics are thought to be suitable to talk about on the phone in which places. This is again connected with interpersonal distances, as third-parties create a feeling of discomfort when they are able to eavesdrop on conversations about sensitive topics. Pictures of four places were shown to the interviewees and they were asked to assess what kind of topics they would talk about in which places.

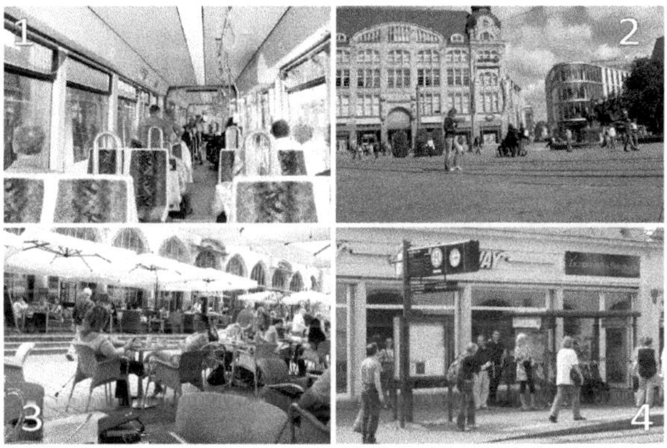

Illustration: Places of public communication: Trams, squares, street cafés and tram stops.

To show the results in detail would go too far. One can say however, that overall there was a pronounced 'Sense of Place' (Höflich 2005a), meaning that people had a sense for which kinds of places are appropriate for mobile phone conversations and which are not. Public places offer a good opportunity, especially when meetings have to be arranged. There have been almost no gender-specific differences. Even though public settings seem to be predominantly male characterized (cf. i.e. Mühlen Achs 2003: 202 et seq.), our inquiries did not show a gender-specific formation of enclaves. There are however gender-specific differences regarding places which to call from. Just one example: Women rather pass their time by making mobile phone calls than men when waiting at tram stops (see illustration).

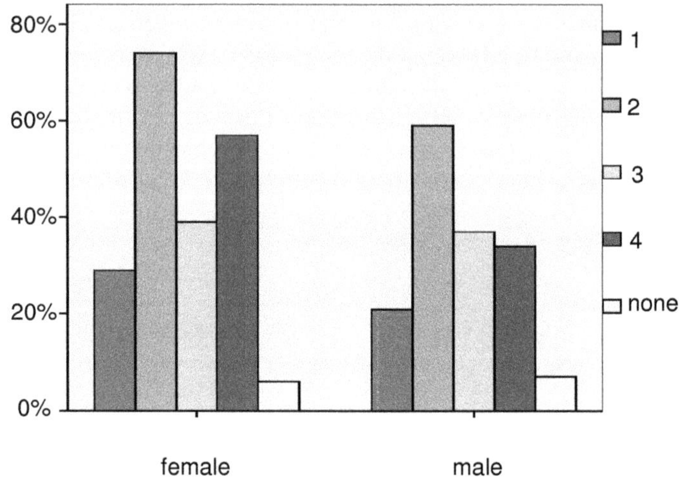

Illustration: Passing time via mobile phone calls

In “mobility gates”, which include situations of waiting, transport and transition and also waiting at tram stops, the risk of being perceived as a nuisance because of making mobile phones calls is rather low. The fact that women tend to make calls more often than men when waiting at tram stops is furthermore not specifically caused by the distinctiveness of this situation, but rather due to the fact that women use public transport a lot more often than men (which tend to use cars more). This is even more so the case in East Germany (Bundesamt für Bauwesen und Raumordnung 2007: 9 / The Federal Office for Building and Regional Planning). In order to be able to analyse gender-specific moments of mobile phone usage, it has to be taken into consideration where people – here men and women – spend their time in the city; how they are embedded into the activity patterns of a city. Especially in the field of feministic urbanism research it is often stated that women – having to link a lot more places with each other during the course of a day than men – travel a lot more in their everyday lives (Löw 2001: 249). One only has to bear in mind the distances that have to be covered on their ways to work, to schools, to grocery stores, to nursery schools etc. This again refers to a gender-specific adoption of space whereby distance behaviour plays a role, but is only one aspect of it. In the end, there is still a substantial amount of research to be done in the field of mobile phone use in public settings.

Conclusions and outlook

New media have to find their place in everyday life at first. What you do with them in the end is preformed by their technology, but in no sense determined. There is, like Pinch and Bijker (1987) state, an "interpretive flexibility" which is more clearly defined once you start using the medium and ends in a "closure", which is the determination or standardization of ways of usage respectively, although this is not necessarily fixed once and for all. Every new medium creates an eruption at first, which shakes up the communicative run of events, meaning the established communicative practices. A standardization only happens after a process of adoption, or domestication respectively (cf. Höflich 2008). Practices have been adapted, but not necessarily fundamentally changed. Although new media create new practices, which furthermore change how we use previous media, they have always also been incorporated into old practices. This also holds true, not only but especially, for mobile phones. Rich Ling (2004: 143) assumes a standardization:

> "The ringing of a home telephone is handled with well-oiled routines – if not excitement in the case of lovelorn teens. Switching on a light is only occasionally the focus of social ritual, as, for example, when lighting the family Christmas tree or dimming the lights for a "special dinner". The mobile phone may, in all likelihood, go the same way."

Such standardization does not only mean a de-problematization, but also an embedding into everyday life. Even though mobile phones are (increasingly) used in the space of our homes, the crux is that you can take it with you and make calls from wherever you are. It is therefore of interest how mobile phones are embedded into certain environments or where they are even banned from. And in every environment there will initially be a dispute over which kind of behaviour is appropriate and which is not. A new medium in public settings simultaneously changes a public order of interactions. The pedestrians' patterns of movement and standstill change for example, one just has to think of the quasi-autistic walking about of mobile phone callers in order to mark their temporarily private space, or the looking for niches in order to be able to make calls without being interrupted by other people or without annoying other people themselves. All of this has to be accepted at first, or respectively in its new shape. This reminds one of the early years of the automobile, which had to assert itself against pedestrians at first. Before cars entered the city, the city belonged to the people, who moved around like they wanted to. Plus, cars were foreign bodies, and dangerous ones at that. Norton (2008) skillfully described this in his knowledgeable view into the early years of automobile traffic in American cities:

> "Beneath the grief and anger of many safety reformers lay an old assumption: city streets, like city parks, were public spaces. Anyone could use them provided they did not

annoy or endanger others. Under this construction of the city street, even children at play could be legitimate street users, and even careful motorists were under suspicion. In the 1920s, however, the pressure of traffic casualties divided old allies. Some renewed their resolve to compel motorists to conform to the customs of the street as it had been, especially by limiting their speed. Others, more pragmatic, wanted to save lives by giving pedestrians more responsibility for their own safety. Finally, some newcomers proposed a more radical social reconstruction of the street as a motor thoroughfare, confining pedestrians to crossings and sidewalks" (Norton 2008: 65).

This shows parallels to mobile phones or mobile media in general. Just like some city councils are thinking about a ban on iPods on all streets, they also thought about banning the public reading of books before – as this, they argue, is a danger to public safety.

People arranged themselves, or are at least doing so now step-by-step. Overall, the results of the study suggest that movement during mobile communication in public settings it not a disturbance (anymore), or is respectively not perceived as one that much. The observations, as well as the descriptions during the group discussions, clearly show that mobile communication now is an everyday part of actions in public settings. The potential for disturbance is thereby not bigger or smaller as with other movements in public settings. Especially the fact, that most people recognize themselves in both roles – the role of the mobile communicator and of the pedestrian – can explain why there seem to be certain arrangements which again guarantee a smooth run of events. The focus on open, multi-functional places however, should not lead to assumptions on mobile communication in (semi-) public places in general. Especially the survey showed that places which are designed for a certain purpose, for example cafés and shops, require special arrangements. It is hereby of notice, that mobile communication can occupy a room acoustically (cf. also Ling 2004: 128), which is an argument that does not seem to apply to open situations which are characterized by movement.

Furthermore there was a trend towards the fact that gender seems to play a role when it comes to occupying public spaces, as well as and related to that, the use of mobile phones. This concerns patterns of activity on the one hand, and distance behaviour as moment of conserving the private in the public on the other hand. Within the framework of the study it was shown, that in regard to the observation studies, evidences relating to that are not that empirically obvious all the time. While the observations rather brought indications forward, the results of the surveys were much clearer. This can be ascribed to methodical artefacts, but also to the fact that gender is reflected in complex interaction connections, which are again characterized by other influences. The relationship between gender and mobile phones consequently has to be researched within the framework of complex social contexts. Even though there are no differences between genders, there is no reason why especially this aspect should not be researched into more detail (cf. further Fortunati 2009). Although the results of the study show that there has been a certain standardization in regard to mobile phone usage in public settings, this does not mean that research on this topic has come to an end. Media use is a dynamic affair, which depends on social influences as well as on the use of other media. This will never rest, as media use is linked with its ongoing adaptation and re-invention.

References

Altman, Irwin (1975): The Environment and Social Behavior. Privacy, Personal Space, Territory, Growding. Monterey, CA: Brooks / Cole Publishing Company.

Altman, Irwin / Chemers, Martin M. (1984): Culture and Environment. Montery, CA: Brooks / Cole Publishing Company.

Anderson, Elija (1990): Streetwise. Race, Class and Change in an Urban Community. Chicago / London: University of Chicago Press.

Bahrdt, Hans Paul (1969): Die moderne Großstadt. Soziologische Überlegungen zum Städtebau. Hamburg: Christian Wegner Verlag.

Barker, Roger G. (1968): Ecological Psychology: Concepts and methods for studying the environment of human behaviour. Palo Alto, CA: Stanford University Press.

Behme, Rolf et al. (1988): Telefonzelle. Flüchtiger Ort der Worte. Dortmund: Schack Verlag.

Bell, Allan (1984): Language Style as Audience Design. In: Language in Society, 13, 145-204.

Bundesamt für Bauwesen und Raumordnung (2007): Frauen – Männer – Räume. Berichte Volume 26. Kurzfasssung. Bonn.

Burkart, Günter (2000): Das Mobiltelefon und die Veränderung der Kommunikation im sozialen Raum. In: Ulrich Beck (eds.): Soziale Welt. Zeitschrift für sozialwissenschaftliche Forschung und Praxis. No 51, Volume 2. Baden-Baden: Nomos Verlag, 209-227.

Burkart, Günter (2007): Handymania. Wie das Mobiltelefon unser Leben verändert hat. Frankfurt am Main / New York: Campus.

Collier, John / Collier, Malcolm (2004): Visual anthropology. Photography as a research method. Albuquerque: University of New Mexico Press.

Dreher, Michael / Dreher, Eva (1995): Gruppendiskussionsverfahren. In: Flick, Uwe et al. (eds.): Handbuch qualitative Sozialforschung. Beltz: Weinheim, 186-188.

Duke, Marshall P. / Nowicki, Stephen (1972): A New Measure and Social Learning Model for Interpersonal Distance. In: Journal of Experimental Research in Personality, 6, 119-132.

Fortunati, Leopoldina (2003): The Mobile phone and self-presentation. Unpublished paper from the 'Front stage / Back stage: Mobile communication and the renegotiation of the social sphere' conference. Grimstad, Norway, 22-24 June.

Fortunati, Leopoldina (2009): Gender and the Mobile Phone. In: Goggin, Gerard / Hjorth, Larissa (eds.): Mobile Technologies: From Telecommunications to Media. London: Routledge, 23-34.

Gergen, Kenneth J. (2002): The challenge of absent present. In: Katz, James E. / Aakhus, Mark (eds.): Perpetual Contact. Cambridge: University Press, 227-241.

Glaser, Barney G. / Strauss, Anselm L. (1967): The discovery of grounded theory: strategies for qualitative research. New York: Aldine.

Goffman, Erving (1966): Behavior in public places: notes on the social organisation of gatherings. New York et al: Free Press.

Goffman, Erving (1971): Relations in Public: Micro-Studies of the Public Order. New York et al: Allen Lane.

Goffman, Erving (1994): Interaktionsrituale: über Verhalten in direkter Kommunikation. Frankfurt am Main: Suhrkamp.

Hall, Edward T. (1969): The Hidden Dimension. Garden City, New York: Anchor Books.

Hampton, Keith N. / Livio, Oren / Sessions, Lauren (2009): The Social Life of Wireless Urban Spaces: Internet Use, Social Networks, and the Public Realm. In: Journal of Communication (in press).

Häussermann, Hartmut / Siegel, Walter (2004): Stadtsoziologie. Eine Einführung. Franfurt / New York: Campus.

Henley, Nancy M: (1977): Body Politicy. Power, Sex, and Nonverbal Communication. Englewood Cliffs, NJ: Prentice Hall.

Höflich, Joachim R. (2005a): The Mobile Phone and the Dynamic between Private and Public Communication: Results of an International Exploratory Study. In. Glotz, Peter / Bertschi, Stefan / Locke, Chris (eds.): Thumb Culture. The Meaning of Mobile Phones for Society. Bielefeld: Transkript, 123-135.

Höflich, Joachim R. (2005b): A Certain Sense of Place. In: Nyirí, Kristóf (ed.): A Sense of Place. Wien: Passagen Verlag, 159-168.

Höflich, Joachim R. (2006): Places of Life – Places of Communication. In: Höflich, Joachim R. / Hartmann, Maren (eds.): Mobile Communication in Everyday Life: Ethnographic views, Observations and Reflections. Berlin: Frank&Timme, 19-51.

Höflich, Joachim R. (2008): Reinventing the Old? New Communication Technologies and Changing Practices of Communication. In: The Role of New Technologies in Global Societies. Theoretical Reflections, Practical Concerns, and its Implications for China. 30th -31st July, Department of

Applied Social Sciences, The Hong Kong Polytechnic University. Conference Proceedings, Hong Kong, 373-382.
Höflich, Joachim R. (2009): Mobile Phone Calls and Emotional Stress. In: Fortunati, Leopoldina / Vincent, Jane (eds.): Electronic Emotions. The Mediation of Emotions via Information and Communication Technologies. New York et al.: Peter Lang, 63-83.
Höflich, Joachim R. / Hartmann, Maren (2007): Grenzverschiebungen – Mobile Kommunikation im Spannungsfeld von öffentlichen und privaten Sphären. In: Röser, Jutta (ed.): MedienAlltag. Domestizierungsprozesse alter und neuer Medien. Wiesbaden: VS Verlag, 211-221.
Humprheys, Lee (2006): Photos and Fieldwork: Capturing Norms for Mobile Phone Use in the US. In: Höflich, Joachim R. / Hartmann, Maren (eds.): Mobile Communication in Everyday Life: Ethnographic views, Observations and Reflections. Berlin: Frank&Timme, 55-78.
Krotz, Friedrich (2007): Mediatisierung: Fallstudien zum Wandel von Kommunikation. Wiesbaden: Verlag für Sozialwissenschaften.
Kruse, Lenelis (1996): Raum und Bewegung. In: Kruse, Lenelis / Graumann, Carl Friedrich / Lantermann, Ernst-Dieter (eds.): Ökologische Psychologie. Ein Handbuch in Schlüsselbegriffen. Studienausgaben. Weinheim: Beltz: 313-324.
Lasén, Amparo (2006): How to be in two Places at the same time? Mobile phone use in public places. In: Höflich, Joachim R. / Hartmann, Maren (eds.): Mobile Communication in Everyday Life: Ethnographic views, Observations and Reflections. Berlin: Frank&Timme, 227-251.
Ling, Rich (1998): "One can talk about common manners!" The use of mobile telephones in inappropriate situations. In: Telektronikk. No.94, 65-76.
Ling, Rich (2002): The Social Juxtaposition of Mobile Telephone Conversations and Public Spaces. The Social and Cultural Impact / Meaning of Mobile Communication. Chunchon Conference on Mobile Communication, July 13-14, 2002.
Ling, Rich (2004): The Mobile Connection. San Francisco: Morgan Kaufmann.
Lofland, Lyn H. (1998): The Public Realm. Exploring the City's Quintessential Social Territory. New York: Aladiny de Gruyter.
Löw, Martina (2001): Raumsoziologie. Frankfurt am Main: Suhrkamp.
Luhmann, Niklas (1984): Soziale Systeme: Grundriß einer allgemeinen Theorie. Frankfurt am Main: Suhrkamp.
Mühlen Achs, Gitta (1993): Wie Katz und Hund. Die Körpersprache der Geschlechter. München: Frauenoffensive.
Mühlen Achs, Gitta (2003): Wer führt? Körpersprache und die Ordnung der Geschlechter. München: Frauenoffensive.
Murtagh, Ged M. (2002): Seeing the "Rules": Preliminary Observations of Action, Interaction and Mobile Phone Use. In: Brown, Barry et al. (eds.):

Wireless world: social and interactional aspects of the mobile age. London et al: Springer, 81-91.
Norton, Peter D. (2008): Fighting Traffic. The Dawn of the Motor Age in the American City. Cambridge, London: MIT Press.
Okabe, Daisuke / Ito, Mizuko (2002): Keitai in Public Transportation. In: Ito, Mizuko / Okabe, Daisuke / Matsuda, Misa (eds.): Personal, Portable, Pedestrian. Mobile Phones in Japanese Life. Cambridge, MA: MIT Press, 205 -217.
Pinch, Trevor / Bijker, Wiebe E. (1987): The Social Construction of Facts and Artefacts or How the Sociology of Science and the Sociology of Technology Might Benefit Each Other. In: Bijker, Wiebe E. / Hughes, Thomas R. / Pinch, Trevor (eds.): The Social Construction of Technological Systems. Cambridge, London: MIT Press, 55-64.
Roeder, Uta-Regina (2003): Selbstkonstruktion und interpersonale Distanz. Dissertation, Fachbereich Erziehungswissenschaft und Psychologie der Freie Universität Berlin.
Rössler, Beate (2001): Der Wert des Privaten. Frankfurt am Main: Suhrkamp.
Schubert, Herbert (2000): Städtischer Raum und Verhalten. Opladen: Leske+Budrich.
Schulz-Gambert, Jürgen (1996): Persönlicher Raum. In: Kruse, Lenelis / Graumann, Carl-Friedrich / Lantermann, Ernst-Dieter (eds.): Ökologische Psychologie. Ein Handbuch in Schlüsselbegriffen. Studienausgabe. Weinheim: Beltz, 325-333.
Schütz, Alfred / Parsons, Talcott (1977): Zur Theorie sozialen Handelns. Ein Briefwechsel. Frankfurt am Main: Suhrkamp Verlag
Sennett, Richard (1974): The Fall of Public Man. New York: Alfred A. Knopf.
Simmel, Georg (1903): Die Großstädte und das Geistesleben. In: Petermann, Theodor (ed.): Die Großstadt. Vorträge und Aufsätze zur Städteausstellung. Dresden: Zahn & Jaensch, 185-206.
Simmel, Georg (1950): The Metropolis and Mental Life. In: Wolff, Kurt (ed.): The Sociology of Georg Simmel. Glencoe: Free Press, 409-424.
Weil, Felix (2005): Privatsphäre – schützenswert und uncool? In: Grimme, Petra / Capurro, Rafael (eds.): Tugenden der Medienkultur. Zu Sinn und Sinnverlust tugendhaften Handelns in der medialen Kommunikation. Alzey: Franz Steiner Verlag, 107-119.
Weiß, Ralph (2002): Vom gewandelten Sinn für das Private. In: Weiß, Ralph / Groebel, Jo (eds.): Privatheit im öffentlichen Raum. Medienhandeln zwischen Individualisierung und Entgrenzung. Opladen: Leske+Budrich, 27-87.
Weiß, Ralph / Groebel, Jo (eds.) (2002): Privatheit im öffentlichen Raum. Medienhandeln zwischen Individualisierung und Entgrenzung. Opladen: Leske+Budrich.

Whyte, Willian H. (1988): City. Rediscovering the Center. New York et al.: Achor Books.
Wolff, Michael (1973): Notes on the Behavior of Pedestrians. In: Birenbaum, Arnold / Sagarin, Edward (eds.): People in Places. The Sociology of the Familiar. New York: Praeger, 35-48.
Wood, Julia T. (2001): Gendered Lives. Communication, Gender, and Culture. 4th ed., Belmont, CA: Wadsworth.

Interaction and individual patterns of mobile communication in public places

Isabel Schlote & Christine Linke

Introduction

This study[1] examines communicative actions of individuals in public space and analyses media-induced usage structures, based on the assumption of progressing media expansion in temporal, structural and social regards (cf. Krotz 2001).

There has been a wide range of thematically and methodically differently conducted studies in the field of mobile communication, which was also found out to be the main form of usage during the empirical stage of this study (cf. i.e. Burkart 2007; Campbell / Ling 2009; Glotz / Bertschi / Locke 2005; Höflich 2006, 2007, 2008; Höflich / Hartmann 2007; Katz 2006, 2008; Ling 2008). But still Katz (2006: 66) calls for a stronger consideration of the desideratum of public space in future studies on mobile communication. This study answers this call. Another attraction that a study of public space provides is the examination of the continuing process of taring between public and private. This topic has come back into focus again with the introduction of mobile phones into our everyday lives (cf. also Höflich 2005a). The public space is an area which individuals use and, simultaneously, shape on a daily basis. Studying communicative actions in public space also offers an opportunity to analyse and capture the use of media as a continuing course of action of people and in how far new media methods of access change social norms and coexistence.

The study approaches this topic on a situational as well as on an individual level and raises the question which forms of media-based interaction can be found in public spaces. Situations offer the frame for communicative actions and refer to social norms and possible forms of arrangement. Not only special features of a room can influence a situation, but also past and future, planned actions of the people involved. Social relationships which individuals maintain with each other and which have different degrees of intimacy thus shape current encounters. Only a combination of interaction-specific aspects and moments of

1 The research has been carried out as part of the research project "Mobile communication, telematization of everyday life and the transition of media practices" at the University of Erfurt which was financed by the German Research Foundation (Deutsche Forschungsgemeinschaft). The authors like to thank their colleagues at the University of Erfurt for their comments on an earlier version of this paper.

social relationships enable an individual to develop a repertoire of actions, also with a view to media patterns. Communicative actions, which consist of situational and meta-situational elements, give clues as to how media can be incorporated into situations without having to re-define all the elements of a situation again. Considering the complexity of such an analysis, this study seeks a comprehensive and open approach on the subject, which was achieved by a multi-methodical procedure. In the first part the underlying relevant aspects of this study – interaction, communication, media and social relationships – are introduced providing a theoretical framework for the aspects dealt with in this study. In the second part the used methods and research strategy are described. Finally, the results are discussed and illustrated by examples.

Background: interaction, communication, relationships and media

Humans as social beings depend upon interactions with other humans. Interaction is to be understood as:

> "reciprocal social action of two or more people, whereby each partner relies in their actions upon the fact that the other partner relates their actions to past or future actions of the first partner" (Bahrdt 1996: 121).

Respective intentions of action can be indicated in a verbal or non-verbal way, whereby in everyday life often a mix of these two can be observed. This "co-orientation" (Scherer 1979) on the other partner supports or even only enables interaction between two people. Individuals use different "tools" for structuring interaction (cf. Goffman 1978, 1972). Hereby, individuals present themselves in an encounter and draw a certain picture of their own persona by using different actions such as manners and ways of expression, for example through fashion. If this picture is being confirmed by the respective counterpart, it can be taken as a basis for further interaction. Thereby, reciprocity of indication and acceptance is needed. The acceptance of the interactional partner can hereby be understood as acceptance of the role offered. However, this role is not to be understood as the kind of role Goffman relates in his book "The Presentation of Self in Everyday Life" (Goffman 2002), in which he describes a kind of behaviour of individuals which primarily serves the function of deluding other people[2]. The basic assumption hereby is that individuals only can or possibly only want to show certain aspects of their persona in interactions; namely those which are relevant to the current situation. Acting a certain role is therefore rather to be understood

2 Surely, people do also deliberately choose a role in order to delude other people. These interaction strategies, which occur in many shapes and forms, shall however not be the subject of this study.

as a management of what to show than as an attempt of delusion. The image that individuals create of themselves through their actions in a situation is closely intertwined with this. The image shown will be compared by the interactional partner with the current image she / he has from the other person and will be accepted or, if necessary, upgraded or downgraded. Individuals use certain techniques to cultivate their image in order to either bring their actions in accordance with their current one or to create a specific image (cf. Goffman 1978). In interactional relationships it is important to save your own image as well as the one of your interactional partner[3] in order to not endanger interaction. Many of these image-saving techniques become habitual or standardised actions.

Individuals meet in different situations every day. The fact, that our everyday life is more and more being permeated by and interwoven with mediated offers, is being reflected in our interactions (Krotz 2007). Encounters between people are more and more mediated, meaning there is no physical co-presence. The ongoing mediatization process does not only change situational arrangements but also established interactional strategies. Mobile phones which unite, depending on their technical equipment, for example TV, radio, games console, camera, writing utensil and telephone in one, thus need a different management of attention depending on which option one uses. A special management of situations is particularly needed when acting in a "two frames" situation where real-life and virtual contacts are happening at the same time (Höflich 2005b). The mobile phone, which is a mobile as well as a personal set of media, is reason and element of the ongoing changing process of communicative strategies. In this context the question comes up, in how far mediated forms of communication influence face-to-face[4] encounters and the other way around (cf. also Fortunati 2005). However, face-to-face communication is the first form of exchange an individual learns despite all technological process. Mediated variations of interaction are only added at a later stage of development. Face-to-face encounters can thus still be regarded as the prototype of communicative processes[5] (cf. Berger / Luckmann 2007; Krotz 2007).

That is why face-to-face interaction is used as a basis for analysing mediated forms of interaction. Basic strategies of human behaviour are observable in face-

3 However, there are certain actions which can endanger an image: The innocent action represents an offence which was not intended. A (calculated) insult on the other hand is intended and deliberately used during an interaction. Between both poles is the accidental insult, which the offender uses regardless of its hurtful consequences.

4 We hold on to the term face-to-face communication, even though Fortunati (2005) rightfully observes that in encounters with physical co-presence the whole body with its verbal and non-verbal information signals needs to be considered. The term face-to-face which is used in this essay does not exclude this aspect, but merely refers to the aspect of physical co-presence of interaction partners in contrast to mediated communication.

5 Considering the rapid development of technological process, it is possible, that this might change as well.

to-face encounters. Individuals use this fundamental knowledge to create and define situations of mediated interactions.

(Situational) influences on interaction

Every communicative action is integrated into the constituent and contextual aspects of a situation. Thanks to its technological features mobile phones can be used everywhere, for example for communicating with physically absent people. This involves an enlargement of context. Depending on the different context elements, integrating mobile devices can be more or less difficult. The evaluation of new areas of usage and the possible modification of situations are both part of the appropriation of a new medium. Processes of change in interactional routines are most apparent in mobile telephoning[6]. There are different arrangements between communication partners during a conversation on mobile phones. The callers have a certain advantage as they can prepare for the interaction, which is initiated by them. They have enough time to make certain arrangements in their own surroundings (i.e. looking for a convenient place to have their conversation in). The callees on the other hand only have a small timeslot of only a few seconds at their disposal to prepare themselves. When their mobile phones announce an incoming call by ringing or vibrating, they decide whether or not they accept it – that is, if they have sensed these acoustic and / or tactile stimuli at all. Not only deliberately unaccepted calls, but also "missed" calls somewhat force callees to reciprocate. The feeling of having to return missed calls may be due to the fact that mobile phones are not only regarded as mobile devices, but especially as personal devices (Höflich 2001). Unlike with landline phones, there is a very high probability that the called person will gain knowledge of the fact that someone tried to contact them. In an earlier stage of mobile phone acquisition there have been indications for a need to explain unaccepted offers of interaction. In this respect, the question arises of whether there is still this pressure to explain your media unavailability in front of others, or whether meanwhile a certain degree of understanding has been developed; that there are situations in life where telephone calls are inappropriate even though one might be permanently available.

Although every interaction and its corresponding / respective situation should be regarded as unique, individuals create an available repertoire of actions in the course of their socialisation, so that they do not have to re-negotiate similar

6 Apart from phenomena such as the dilemma of availability and control functions (cf. e.g. Burkart 2000), which mainly go back to the telephone function, telephoning in public is the most central enlargement of already established situations. Portable games consoles, messengers and calendars already existed before mobile phones. Telephoning in public however was bound to phone cells. The mobile terminal device as a hybrid media unites the functions of many different media in one.

situations again and again. This repertoire consists of rules, habits, rituals and manners. Behaviour patterns show on the one hand that an individual is a person with certain desired and undesired characteristics (cf. Goffman 1978), and ensure on the other hand the communication with one or several other individuals as a reciprocal understanding through a mechanism of simple feedbacks. Although one can assume that in interactions "animosities are sufficiently suppressed, so that they (Authors' note: the interactional partners) might temporarily forget their caution and actually start to speak" (Goffman 1978: 48 et seq.), incidents may occur in the course of interactions. Thus, the earlier stage of mobile phone use led a time and again to crises. However, acting strategies and rules of use have been developed with rising number of users and a higher degree of appropriation of the medium into our everyday lives. Such rules range from juristically fixed rules of behaviour (i.e. mobile phone use while driving) to commonly accepted regulations. However, even though mobile phones have been incorporated into the existing media ensemble, their use still creates eruptive moments which lead to re-evaluations of situations. Garfinkel's (1984) breaching experiments serve as good examples for what happens when people do not follow existing arrangements anymore. These impressively illustrate versed handling of everyday situations[7] and show at the same time individual's strategies of normalization in critical situations. Rules of behaviour and everyday practiced behaviour patterns can only be reflected upon and described theoretically, when they are not taken for granted on an everyday level. Apart from different definitions of certain situations, there are also unintentional incidents when individuals come into contact with each other. In these cases, the individual has a set of reparation measures at hand. Interactional partners can for their part overlook such a faux pas with "polite indifference" (Goffman 1971: 150). However, if the cut into the interaction is too substantial, the "offenders" need to take corrective actions. In most cases they are being given the chance to explain their misdeeds, or to make up for them, thus restoring proper order (i.e. by offering an apology). If the interactional partners are interested in maintaining the interactional relationship they will accept the offer, as long as they think it is authentic. The "offenders" on the other hand should signal their gratitude in case of acceptance (cf. Goffman 1978)[8]. Non-verbal signs gain importance in interactional relationships during encounters of

7 In one experiment participants answered to the question of "How are you?" with "How am I in regard to what? My health, my finances, (...)" (Garfinkel 1984:44) and thus gave an everyday commonplace phrase a deeper meaning which was not anticipated by the interactional partner. Similar confusions would occur were a sales person not accepting the medium of money as means of payment for the purchase of a certain product (cf. also the study of Höflich (2009) on normalization strategies in mobile communication).

8 The steps of action shown may vary and nevertheless lead to a successful correction of the faux pas in an interaction.

people who are not acquainted with each other. People fall back into this communication channel to signal "non-aggressiveness" (Goffman 1982: 112), for example by smiling at someone after bumping into them. They want to show that their actions were unintentional and at the same time restore normality.

Furthermore an analysis mediated interaction in public space reveals that there is a trend to relocate a previously private practise into public space (cf. Höflich / Kircher, in this volume). When talking about public space, we mean the front stage as described by Goffman in his early dramaturgical approach. Individuals have access to this stage and know that other individuals – known and unknown – are aware of each other[9]. There is only a thin, blurry line between both. Mobile phones provide the technical requirements to negotiate private matters in public. At the end however, it is the people themselves who relocate this privacy into public space, the mobile phone is merely the artefact which enables them to perform this process. In this context, one should be reminded of the fact, that the border between private and public sphere has always been a flexible one, even before the proliferation of mobile phones. This process does not only work in one way, meaning from private to public, but also vice versa, as when people began to bring the public into their private spheres by buying television sets. To assume a strict border, or even the removal of one in favour of a tyranny of intimacy (Sennett 1986) enabled by the new media communication technologies, would mean to neglect a thorough analysis of the media ensemble as a whole and with its specific characteristics.

Furthermore, there are also different areas within public space. These suggest a particular set of actions or come with a special set of rules and regulations (cf. also Burkart 2000). The latter one can be of a general or specific nature (cf. also Klamt 2007). In restaurants for example, as well as in public space in general, one is expected to dress appropriately. However, while bringing your own food may be convenient when travelling by train or when resting on a public square, it would be considered as a breach of rules in restaurants. There are also certain rules of room use in restaurants or cafés. Individuals claim a table for the duration of their stay. It is mutually agreed upon that without good reason other people are not allowed to claim access to this territory[10]. Therefore this creates a connection between private and public space. A restaurant or café is furthermore a good place for getting into contact and communicating with other people.

---

9 Depending on the relationship a person is allowed to perceive / to notice more or less about a person. A brief glance to another person on the sidewalk is accepted or even desired in order to avoid collisions. Staring at an unknown person however, is perceived as too much attention being given.

10 If a guest were staring at the plate of another person, even if just to look at the chosen meal, then this action may be defined as well as an incursion into the presently claimed territory.

Taking up again the aspect that people use their situational experiences to define new situations this paragraph shows that this process also includes some problematical aspects, e.g. if a known form of usage as having a consversation via telephone is taken to a new context that may not offer the preconditions for such an action. In this case people had to learn and find arrangements how to integrate the known action into a situation that oftentimes is also known but not as a frame for this special behaviour. That does not mean that a certain form of media usage automatically is integrated in a specific context and it only takes some time. A possible arrangement may also be that the media is forbidden or simply not used in the context. It is necessary to also incorporate the specific requirements of a room into a model for mobile communication situations.

Mediatized interaction

The described routines of action can also be found in media-supported interactional processes. The re-regulation of attendance becomes a central point of analysis in the changing processes which are induced by new media offers. Individuals have to pay attention to two synchronous situations[11] (cf. Höflich 2003), as one is "physically present but is absorbed by a technologically mediated world of elsewhere" (Gergen 2002: 227). Meanwhile, specific arrangements for mobile phone use in the presence of other people have been established, whereby different combinations of media and face-to-face communication can be observed. If one regards the described aspects as parts of a media-supported communication model than potential third persons have to be added to this model (cf. Höflich 2005b: 28). Especially focused interactions (Goffman 1971) call for a swift coordination process of the callee as they want to handle the approaching situational interconnection without any critical incidents happening; thus confirming or preserving the current type of relationship to the respective partner. The incoming call which interrupts face-to-face communication, and the following process of negotiation which serves as a demonstration of normality, do not necessarily have to be observed by the communication partners as an eruptive or interaction endangering moment. This also depends on how far the communication partners have integrated the ability

11 Certainly, this is an already-known interaction strategy, which is for example being used when an individual (A) is moving in public with an acquaintance and meets a third person. Both people, who each have an own relationship with A may now know each other. For the time of the encounter an arrangement is needed in order to handle this double interaction and treat each interactional with due respect. Hereby, A certainly depends on the co-operation of his interaction partners, for example in the shape that A's companion temporarily releases him / her from the interaction so that she / he may get into contact with the other person. In turn, the interaction with the incidentally met person will only last for a short period of time.

to switch between face-to-face and media situation in their numerous facets into their repertoire of possible courses of action. Generally, there is no direct negotiation process with unknown / unacquainted present people in public space. Nonetheless, telephone users have to split their attention, so as to avoid critical incidents, like they have to with every other action in public space (i.e. avoiding bumping into other people). Hereby commonly used rules of interaction are mostly being used. Apart from the special situational requirements individuals have to integrate the mobile phone firstly in their own media repertoire and secondly have to come to an agreement with their interaction partner in which way the media should be used. Here one should also keep in mind that an arrangement is needed for face-to-face encounter and the media usage during these encounters, e.g if it is appropriate to answer a call while having a conversation with somebody face-to-face. In this context the kind of relationship people have as well as the activity pattern of a person are of importance.

The role of social relationships in interaction

As already mentioned, interpersonal communication is embedded into everyday life and therefore structured by daily courses of action (Wood / Duck 2006). Everyday structures of people can be described as activity patterns, and therefore temporal and spatial aspects should be taken into consideration. The behaviour of an individual is of interest when studying "Human Activity Patterns". Based on that, Chapin (1974) observed the behaviour patterns of urban metropolis populations on an aggregated level and describes activity patterns of city residents – their ways of allocating time in the course of a day to different activities, the rhythm of these activities around the clock and the places of action of these activities within the realms of the city. "Human activity" consists of physiologically restricted and learnt behaviour patterns. "Activity variations" are determined by seasonal aspects and also by variations between years, which reflect changes within the individual's daily rhythm. The characteristics of physiologically determined activities basically dictate the chronological rhythm of the individual's routines and activities and thus affect the chronological order and structure of all other activities (Chapin 1974). Naturally, the individual's activities have to be regarded within the framework of its social embedding, its relationships and way of life.

Relationships between people have a fundamental effect on communication and are always present (Duck 2007). One differentiates between more formal, functional, impersonal relationships and personal or even intimate relationships between people (Miller / Steinberg 1975). Many relationships are based upon role expectations and personality-specific moments. Role relationships (i.e. between traveller and conductor or cashier and customer) consist primarily of culturally determined or organisation-specific expectations on interactional behaviour (Asendorpf / Banse 2000). For interpersonal communication between two people this means that the different interactional patterns, which may also develop in the shape of habits, rituals, patterns and arrangements (also for mediatized patterns), can be shaped on a relationship-specific, on a relationship-unspecific (socially and culturally influenced) and on a role-based level. The level of relationship specificity is thus more important than general role relationships. Another distinction of social relationships can be made between symmetrical and asymmetrical respectively complementary relationships. A symmetrical relationship is characterized by a similar manner of expression by the parties involved, which again expresses itself in a similar way of communication where all parties involved have equal rights. Asymmetrical relationships on the other hand are based on differences with an exchange of complementary actions. This kind of relationship can be found in hierarchies of group systems, such as between teachers and students or superior and employee. The communication of these relationships is (partly) linked with specific demands. However, mixed varieties of these two types of relationships are also conceivable within the structuring of the actual communication between people. When people label a relationship and classify themselves as being in this type of relationship, they have specific expectations on it (Knapp / Vangelisti 2005). Different types of relationships, such as between sibling, between friends or between husband and wife, could be established in everyday life, because these expectations are shared by a large proportion of the people.

An interactional story develops in relationships which in turn always influences the current interactional episode. A few interaction episodes again create a solid interactional pattern. In layman's terms, one would now speak of a relationship between two people which other observers can classify as such as well. In personal relationships a set of different behaviours during interactions is being created (for example for different surroundings, topics, emotions). Consequently, expectations on the relationship and the course of communication with each other are being created, and predictions on future behaviour are being made. So-called interactional scripts are being developed, which act as screenplays for one's behaviour during exchanges with other people (Asendorpf / Banse 2000). Kruse uses the term of "script" in relation with the concept of Behaviour Setting (Barker 1968) and characterises it as a cognitive representation, as a plan on how people behave and communicate in specific situations (Kruse 1986).

Communication constitutes a relationship and simultaneously indicates this relationship to third parties, as the way people communicate with each other conveys the nature of their relationship (Duck 2007). In our increasingly mediatized everyday life, in which relationships are being embedded, media use is linked with the type of relationship people have with each other (Höflich /

Linke forthcoming). Different types of relationships are characterised by a different use of media or in a special use of a medium. At the end, the importance of a medium depends on the type of relationship in which it is being used. Appropriation and use of communication media are social processes in which the influences of media use on the structuring of communication gain importance. Communication media thus influence relationships in a recursive sense. The proliferation of mobile media is an important reason for the establishment of a new quality of communication patterns and uses (i.e. Ling 2008; Höflich / Linke forthcoming).

The use of the mobile phone in public spaces has become integrated in people's everyday practice. A deeper analysis of this practice requires, however, a consideration of situational as well as individual aspects of everyday life. For answering the question on media-based interaction strategies in public spaces the context and especially the specific characteristics of the places, the type of relationship that create certain forms of interaction have to be considered.

Method

An observation study (cf. basics on observational methods, Angrosino 2007; Gehrau 1997; Lamnek 1995, 2005b) carried out in spring and summer 2006 was the first approach to the field. The observations have been conducted in different places. The places had a specific degree of structuring that means ranged from a multifunctional public square in the inner city of the German town of Erfurt, to train stations, bus stops and means of transport as "mobility gates" ("Mobilitätsschleusen") to cafés which suggest certain patterns of behaviour (cf. Burkart 2000). In this first observation cycle the use of mobile phones had been observed exclusively. These observations provided an appraisal how and to what extent the mobile phone is used in the different places. The observations were written down by hand and transcribed in digital form afterwards. An observation protocol had a set header in which information such as date, time and weather, but also the person's age and gender were written down. The record of the actual observation was subsequently written down as a continuous text. Memos on particularities, and photos of the different places completed the observational unit. The collection and analyses of the data was done synchronously. A team of researcher edited the observations and figured out categories. The data analysis is orientated at the principles of Grounded Theory (Corbin / Strauss 1996; Mey / Mruck 2007). The categories were compared with each other. By doing so the phenomena were not only describable but also comprehensible theoretically. Another series of observations followed in 2007, which however had a few methodical changes. The aim was to gain a more complete rendering of the whole situation of mobile phone use. That is why two observers each wrote a protocol for the same sequence of action. One person wrote the events down as usual, and the other researcher recorded his thoughts on a dictaphone. The realisation of this procedure was not always easy, as the observers sometimes felt that they became the centre of attention of other people present, and were in turn observed by them. This aspect has been reflected upon during the talks of the researcher group. Thus, it was important to create adequate areas of protection in order to conduct the observation. Another methodical advancement was the follow-up-interviews of the people observed. After the observers were done with their mobile phone operations they were asked to give a brief interview. These short questioning-units covered those details, which could not be deducted by observation alone, such as the reason of the call. In this context, the observed people were also asked to consent the use of the pictures which were being taken of them during their media use. All mobile phone users which agreed to do an interview, also agreed to the use of their pictures. The limitations of this observation method became clear when compared with the data which has been gained by the interviews. At the same time it also showed all the advantages of an empirical, reality-picturing procedure, which can be used regardless of the persons observed (cf. also Lamnek 1995). A lot of the observational units could not be completed during the second phase and were thus of no use for the evaluation, as the observed people refused to do the follow-up interview. The observations of the second phase are also available in digital form. The data analysis followed the same criteria as mentioned above.

The experience, sighting, compression and analysis of the observational data created the basis for the following six focus group discussion (cf. Barbour 2007; Lamnek 1995, 2005a; Morgan 1997). Each discussion lasted between 60 and 90 minutes and group size varied between three and six people. Although the topic of mobile phone use is not a very emotional one, which might induce a high level of consternation, we deliberately chose a smaller than usual group size. By doing so, we wanted to enhance the special characteristic of this method, the data collection through interaction (cf. also Morgan 1997), as well as achieve an as detailed as possible description of the topic (cf. also on the topic of group size and interactional dynamics, Lamnek 1995, 2005b). Regarding age, the groups have been chosen to be homogeneously, although it has to be said that the age structure differed from group to group (with participants ranging from 16 to 65 years). This procedure seemed sensible with regard to different media socialisations, as it was not the primary purpose of this study to create media biographies and show differences between generations, but to encourage a subjective assessment of media-supported communication processes in public space. The generation issue had been introduced and reflected upon by the participants themselves and a cross analysis of the data also indicated age related differences, which again created a possible basis for further studies. Except for one group, the participants of the group discussions knew each other from school, university or shared courses. This fact did not have any tangible negative effects on the discussion, as there seemed to be a certain level of acquaintance among the participants from the start. Furthermore, the participants used each other in order to contrast or support their own points of view. They discussed controversially and reflected upon their own appropriation of mobile phone use by using the statements of others as a basis. The moderator of the group discussion used a guideline which has been developed by using the results of the observations and the questions which arose from them. This guidelinewas complemented with visual stimuli in the form of pictures and videos (cf. also Banks 2007). These staged observed situations served as a visual support, when certain situations were referred to as the basis for a following discussion. Thus, a common starting point for the discussion was ensured for everyone. The participants reacted positively to the shown material, as they referred time and again to certain staged situations during their argumentations. The moderator of the group discussion mostly stayed in the background during the talks, only giving new impulses when necessary and leading the discussion back to the main topic, when participants began to focus too much on irrelevant side lines. However, it was taken care to incorporate the participant's aspects and to give them time to develop thoughts and lines of argumentation. The remarks on anticipation of situations, as well as self and external assessment of mobile phone use by the participants were of special interest during these talks. The talks have been recorded and put into writing in consideration of simple rules of transcription (cf. Kowal / O'Connell 2005). The group discussion were also analysed taking the research process of Grounded Theory Methodology as an orientation. The categories found in the group discussion were discussed within the research team and in a further step compared to the results of the observations. By doing so, situations of media usage could be described as highly complex.

To summarise: the used method emerged from the antedated observational study. The first observation study concentrated on the use of the mobile phone in public places. In a next step the whole media situation was taken into account. For getting a larger picture of the situation two people accomplished the observation and the observed people were asked some short question afterwards. The guidelines for the group discussion basically emerged from the results of the observation. The group discussion aimed at learning more about how people define and evaluate mediated forms of interaction in public places.

Empirical findings

Central concepts could be worked out during the analysis of the empirical data with regards to interactional and activity structures in public space. The Findings illustrate that mobile phone use takes place in consideration with situational circumstances. Hereby, people anticipated their general own situation and the one of their communication partner as well as the particular requirements of the room. Individuals create a specific management of attention by taking the social relationship with their face-to-face and mediatized communication partners into consideration, which again is represented in respective interactional arrangements. Both aspects mentioned, namely the anticipation of the situation as well as the definition of social relationships, are being reflected in temporally and spatially structured patterns of activity. These findings on mobile communication in public places will be described in detail in the following two paragraphs. While the first one concentrates on the findings on interactional strategies, the second is one presents the results on individual patterns.

Findings I: interactional strategies and mobile communication

A mediatized conversation differs in certain characteristics from face-to-face communication – information has to be communicated verbally, because it is not possible to send non-verbal signals via telephone (unless you have a videophone). Also, as communication partners cannot experience the current situation of their counterparts in a direct way, they have to envisage it in order to align their own actions to the ones of their partners. Although media actions have different interactional demands, not every observed strategy of use has been developed for mobile phones. As already mentioned established routines of actions which have been used with other media offers, for example landline phones, have been adapted for mobile phones as well. Problems occur in these cases where the context demand another course of action, meaning that an established course of action has been taken out of context and put into another one without taking its specific regulations into consideration. Furthermore, the findings show that there are basic expectations for structuring interpersonal cohabitation, even for the use of media in public space. In one of the group discussions a participant stated that "everyone can do what they want", but "have to take others into account. One should not inconvenience other people" (Quintus). Such a concept of behaviour serves as a benchmark for the assessment of other people's intentions of action during encounters.

It was found out, that in public space some accessible, differently regulated areas were more appropriate for certain actions or more specifically, holding conversations on mobile phones, than others (cf. Ling 2005). A group discussion participant summarizes this as follows:

> "It depends also very much on the place, where you are in this moment, when I am sitting in a café and everyone is conducting phone calls, then I have no qualms about doing the same when I am being called if need be. On the other hand, I would be

embarrassed when my phone rings when I am sitting in the front row of the theatre" (Quentin).

It became clear that sometimes, mobile phone users actively look for certain places in which to hold their conversations. At the same time these places only gain their special characteristic by the individuals' ways of use. When two persons are in the same place during a face-to-face communication, they can react to it accordingly, as they both have the same basis from a physical point of view at least. However, during a medially conveyed contact both partners have to clarify their spatial requirements. Most people – no matter if during the short interview after the observation or during the group discussion – had a feeling for respective contexts and their degree of structuring. Based upon this knowledge, the interviewees assessed their possibilities of action (cf. also Höflich 2005c). Furthermore, it was important to the interviewees to know the place and situation of their communication partners in order to structure their calls in one way or the other.

Certain ways of action especially develop between people who know each other very well, and communicate – also via media – on a regular basis; thus creating a common definition of the telephone situation. Conclusions for action are also drawn out of the anticipation on how you want to be treated yourself in the current situation of the medially or physically present opponent. For example, calls are quickly ended as the interviewees would also find it unpleasant if in a café their interactional partners would have a longer conversation on their mobile phones. This is an example of a media-induced situation which can be found quite often these days. A double arrangement has to be made immediately after an incoming call. The contact, which is placed medially, is often – may it be only for some seconds – being given more attention than the partner[12] who is present in that very moment. Here, the anticipation of the caller sets in by asking for the whereabouts ("Where are you"?) or for the availability of attention ("Am I interrupting you in something?") (cf. Höflich 2005b). Furthermore, it was found that in accordance with the answer additional steps follow, which can include a longer conversation, a brief exchange of information or an immediate discontinuation / postponement of communication. Nevertheless, some participants of the group discussion stated that they would reflect over the location or situation of the desired conversation partner prior to dialing the respective number. Such an approach often presumes a fundamental knowledge about the daily routines of the person to be called.

12 It is surely thinkable that a person does not wish for any interruptions when being in a face-to-face encounter and completely remains with his or her attention at the present counterpart and as a consequence lets the telephone ring or chooses a respective setting on the mobile phone, which avoids an interruption.

Besides the anticipation of the prevailing situation and the physically and / or medially present interaction partners as well as one's own presence, the relationships between the individuals refer to potential behaviour patterns[13]. Will it suffice to give unknown present people just as much attention in order to avoid inconvenience, does the direct contact of two acquainted people demand a significant share of their attention (cf. Goffman 1971). Generally, a ringing telephone rushes rather unexpected in a proceeding interaction and promptly demands from its owner the respective action: Either to answer the phone and to withdraw the attention from the established interaction or to remain with this interaction and not to answer the phone. The empirical material reveals that an interruption by mobile phone does not need to be explained to such a great extent to an acquainted person than to a lesser acquainted person. The statement of a participant of a group discussion clarifies with respect to mobile communication the associated scope that goes along with the definition of the relationship:

> "[...] to me it is like that: the better I know the people I am with, the more likely they show an understanding once the telephone rings. Basically, when I am meeting up in a business related context and I am in the middle of a conversation then it is generally for me and my guest more inconvenient being interrupted by phone calls. Whereas in a family-related context it is basically more informal and one can answer the phone and can talk as long as one pleases. People from work generally do not have an understanding for being interrupted by telephone calls, because those people do not have much time. They want you to engage with them and to concentrate on the matter otherwise it is a waste of time and this goes almost never down well" (Florian).

A different evaluation regarding the usage of mobile phones during established interactions was given by Jana. During a face-to-face meeting she never answers a mobile phone call and expects the same from her counterpart no matter which kind of relationship the two of them have and she especially assumes this kind of conduct when the communication was extra arranged. During this time she wants to be "with this person and not with his or her telephone". Despite all the media structuring and acceptance in our everyday life she finds such a conduct "outrageous and rude". If necessary she expresses her request for an unimpaired interaction even to her interaction partners, by asking him / her i.e. "that when meeting up again, then I generally tell him to please not do that again, because I don't want this. Then I don't need to meet up." An exception would be if the interaction partners priorly announce that they expect an important call, which cannot be postponed, thus the possibility of an interruption has been established which means that prior to the reception of that call a situational definition has been made. That is, the interruption may still be present; but, this does not come

13 Of special interest are negotiation processes between people of different status, i.e. the relationship between superiors and ordinary employees. However those data only appeared marginally in our study.

completely unexpected. In this context she further states, that this is a special kind of conduct, which should remain as such, and thus it should not become a standard routine.

At this point one can acknowledge that the results point out that people do have a certain feeling for the situation they are acting in and also anticipate the situation as well as the circumstances of their encounter regardless if it is a face-to-face or mediated form of interaction. It has also become clear that alongside with a general agreement regarding the behaviour in public places each person has their own perspective on which comportment of their encounter he or she is accepting. But also this definition is not a fixed one as people do take into consideration the personal circumstances of their encounter when defining the present co-situation.

Forms of attention management

If there is a medially placed interruption of the interaction respective corrective measures depending on the evaluation of the relationship form and the situational definition of each and every individual are needed (cf. Goffman 1972). Especially interesting is the negotiation process in relationships that are based on a professional context. Various nuances and stages of hierarchy in the relationship have an impact on media interaction. Especially, different hierarchy positions can lead to the fact that these interaction partners can have a greater scope of conduct at their disposal than others. A firm's most senior director can have a conversation on the telephone when meeting up with an inferior employee without having to face negative changes in their relationship definition. Due to a different balance of power such conduct remains unattainable for the employee. This aspect was only seized rudimentarily during the group discussion. Even in the actual observational study, the mobile phone owners were generally accompanied by a person they know as a friend or met an acquainted business partner.

Our Findings suggest that moreover the number of people present leads to various forms of integration of mobile phone use. It is much easier for people to find an agreement if several people are present. The person who is talking on the phone will be released from the face-to-face conversation and is re-integrated into the actual conversation after abandoning that activity. Such a conduct offers the owners of the phones the possibility to divert their full attention to the conversation on the phone, which, however, some participants of the group discussion conceived as unpleasant. A young woman stated with regard to the matter the following:

> "As soon as I answer the mobile phone the communication situation we are in will be interrupted because I communicate with the partner on the phone [...]" (Kathi).

Iwan expressed himself similarly. When dealing with important conversations he prefers to withdraw completely from the group, since the other people do not know the subject of the telephone conversation and he would feel "completely out of place". If the people who remain in this situation still have each other for face-to-face communication then they find themselves in an equivalent situation. With only two people present this is not the case as one finds himself without a respective engagement. Consequently, for the duration of the telephoning activity one communication partner has to find himself another activity. During an observation a young man read the booklet of a CD while his companion was talking on the phone. Although the owners of the mobile phone leave their companions for the duration of the conversation "alone", this does not seem to lead to any considerable disturbances in their interaction relationships. Furthermore, these companions regard the telephone use of their interaction partners as normal and regarded the question whether or not they would feel disturbed by the conduct of their companions as unusual.

Apparently, people feel most uncomfortable when sitting alone in restaurants and cafés. Such a situation cannot be exclusively attributed to the usage of mobile phone, but also arises when the interaction partner is late or goes to the restrooms. At times, in such situations the mobile phone becomes to a kind of "companion substitute" (Goffman describes the same for the newspaper, Goffman 1971: 135) or serves as an evidence for the fact that one has got a social network, that is, "one has not got the feeling of sitting alone here and the people around do not think that one has not got any friends [...]" (Ines). The use of the mobile phone and may it be only the glimpse on the display signals: one is "busy [...], even if you are not" (Florian). The situational definition happens according to the presumptions which are made on account of the presumptions of the other people present. The observations reveal that there are time and again phases in encounters during which no communication happens. For instance, in a café two people were sitting abreast without communicating[14] much at times. The use of the mobile phone in such phases could be arranged rather unproblematic. Moreover the presence of the other person suffices in order to bear non-communicative phases.

The communication on the mobile phone does not necessarily mean the complete withdrawal from the current situation. In case the physically present partners know the communication partners on the phone then the communication can indeed switch to a hybrid of face-to-face and media communication.

> "[...] when the caller is family and you give a sign who it is and the others know the person, then it is a common matter, everybody knows the caller and then it is interesting

14 Hereby, it would be premature to conclude from the observations that there are frictions in the relationship or that these people do not have much to say no one another.

for everyone, because the callee could then switch the phone in the speaker mode [...] and everyone could comment, well there the communication merges into a group communication" (Ingo).

Generally, the concerned parties merely exchange best regards with the caller. However, the ceremony of exchanging best regards alone links both situations. Another way of merging situations is to involve the present communication partner thematically into the conversation, for instance, by asking his / her opinion on a certain aspect. An observation at a public swimming lake in Erfurt refers exemplarily to another kind of media-supported and face-to-face communication merger. A man was talking on the phone and simultaneously interacted with his daughter who took her father's wallet out of a bag. The subsequent interview exposed that this man had talked to his wife. His daughter has got desires and a short explanation suffices in order to explain his action and in order to explain the situation to his wife who cannot observe the situation at the swimming lake.

Place as well as relationship status lead with present contacts as well as with medially placed contacts to different forms of attention management. During a telephone conversation the latter can range from interaction linking, thus the involvement of the physically present conversation partner into the conversation, to an interaction shift, thus the temporal shift of attention for the benefit of the telephone call, and finally to the interruption of the interaction, that is, the temporary physical removal from the face-to-face interaction. Especially the observations indicate that asynchronous forms of communication, like text messaging, are often integrated into the interaction. A young woman was presumably typing a text message on her phone while walking aside of her companion. While typing, the two woman converse. In such situations a respective management of attention is easier, as the asynchronous communication allows for an interruption of the action anytime. That way, a young woman interrupted her typing on the phone time and again in order to communicate with her companions. She prioritized the conversation with her actually present friends over the activity on the phone and she only pursued her telephone activity when there was sufficient attention[15] for her actual face-to-face conversation.

The structuring of interactions between two people who do not know each other also requires a respective management of attention. As already mentioned above, other individuals should be paid as much attention in order to avoid – in this case media-related – incidents. However, the findings suggest that mobile phone users can also "remove" themselves from the situation and thus assign the

15 This arrangement, either the simultaneous or time-lagged typing of text messages, seems to emerge from the media acquisition process. With regard to this matter a pupil explains: "Well, you have learnt it meanwhile.".

responsibility to other people who move about in public space. People, who look down while talking on mobile phones, protect themselves from social contacts with this behaviour since he / she does not offer any possibility for interaction. They create an imaginary box and signals a kind of temporary non-participation (cf. protection-shield function of media Fox 2001). With this strategy individuals trust upon the fact that other people take over the coordination, hoping that no major incidents occur. If mobile phone users bumped into somebody else in such a situation, they still would have some options at their disposal (i.e. a smile) in order to prove his / her "non-aggressiveness" and to avert sanctions.

Sound space and mobile phone talks

The coordinative requirements in public space seem to be known to the individuals from their everyday life. Even though people may also be forced to eavesdrop a face-to-face communication while being in the public the aspect of the intrusion of the mobile phone user into the "sound space" (Goffman 1972: 46) of other people present is regarded as problematic. The evaluation of which matters can be or are allowed to be discussed in public settings is very closely intertwined with the former. The assumption, that adolescents, who were stamped by mobile phones during their socialization, set those boundaries differently was rejected by the empirical findings. Approached upon the aspect of telephoning in public a young woman who was not interacting with her companions while talking on the phone said in the short interview following an observation, it would not be disturbing as long as it is:

> "Nothing private or profound [...]. Meaning, it is only about arranging a meeting or something like that, and then it is okay for me. However, if it were something of a private matter then I would withdraw from the group".

Simultaneously, we found that people are aware of the fact that all publicly discussed matters are accessible for the respective people present. Those are forced to eavesdrop time and again (Ling 2002). Some of the participants of the group discussion admitted on eavesdropping on other people's phone conversation. The people surveyed stated that, to a great extent, they would dislike such behaviour. Public transport and the media behaviour of some travellers were time and again mentioned in this context:

> "Well, I also experienced this when being on business trips. When sitting on a train, in the first class car, then you can witness everything there (-) as everything is being worked off on the phone. You can track everything and (-) Well this is actually not in order [...]" (Frauke).

Admittedly, the participants of the group discussion have a rather low appreciation for such conduct; however they almost never intervened for sanctions. In one case a woman stated that she, from time to time, has got the feeling she has to "call for a halt", when people on the train are talking the whole time about aspects of their private life. Another strategy of hers to stop such telephone calls, she describes hereafter:

> "[...] well, when somebody is constantly talking on the phone on the train, then I say very loudly: "I am not at all interested in that!" Then I say it again and in this moment the people know that I mean them. Well, "I am not at all interested in that" I say and I look at them. And then they realize that I (-) am able to hear everything they say. Often it is just a compartment (-), with eight seats (-) for eight people (-) and then there is one who is talking and talking. And this way I have turned the situation (-) or someone leaves the compartment (-) for the call. Because I think I do want to know about the private life of the other when I am in this private" (Nadine).

Although, it can always be conceived as a nuisance when one has to participate in private conversations of unknown people in public space, one participant pointed out an interesting aspect. The evaluation for the prevailing situation is closely linked to how this woman feels on this respective day, "how sensitive" she is. Depending on her personal shape she assesses a telephone conversation about private matters as more or as less disturbing. She remarks that on some days:

> "I went to them and asked them [...] to keep it down, because I am not necessarily interested in the outcome of the job interview at Frankfurt Airport" (Jana).

Especially, she acts like that when she sees no other option, like retreating herself from this person.

The empirical findings reveal an extension of the repertoire of interaction forms by the additional media component. Such an extension requires the readiness of the individuals to admit new situations in the constituent inventory of situations. If this requirement is not granted the new media could merely be seen as a problem. Ultimately, it is, however, only the element that new situation forms can emerge. However, the latter will only be established if the individual approaches the media in one way or the other, embraces it and develops specific patterns of usage.

Findings II: individual patterns and mobile communication

People's repertoire of behaviour in public space is tied to the specific situation, or even, it arises from the context that the respective person defines (Dourish 2004). However, one has to consider that people implement meta-situational behaviours and patterns. These result from individual activity patterns, which find expression in time and mobility progressions. Both, the activities of the past as well as those, the individual anticipates or plans for its everyday life, are of importance for the individual when defining the current conduct. Thus, mobile communication in public space is observed from the perspective of the individuals' everyday life progressions. It is of particular interest which structures are revealed as well as which influences between individual patterns emerge through mobile communication. By combining the open observation and the adjoining interview, we could shine a light on individual behavioural patterns especially in the second stage of the study.

The aspect of time and individual activity patterns

Observing from this perspective it was established that mobile communication in public space is linked to the interruption, the bridging as well as the combination of people's activities. The findings show that people who used mobile communication in public space, were often situated in a phase between activities (i.e. on their lunch break) or in a period of bridging two activities (i.e. the period between visiting an exhibition and the departure of a train). In some cases the people surveyed were on their way to more or less definite appointments with other people or respectively sojourned for this in that place. And mostly their mobile communication was related to the coordination of a meeting, too. It became clear that the people interviewed classified their behaviour with respect to time and space. For instance, they reported, when asked for an explanation for using their phone, about activities they did beforehand ("I am just coming from work", just have been at the doctor's") or about future activities, and here, once again, the time ("now", "soon", "tomorrow", "on Thursday") as well as the spatial dimension ("over there in a café", "at the school", "here in Erfurt", "at the meeting point") emerged. Weilenmann and Leuchovius (2003: 5) have outlined the intersubjective importance of places in relationship to mobile communication and revealed that people use specific terms for places which are either known ("I'm in school"), relate to already witnessed situations ("where we met last time") or which explain the movement between places and the estimated time of arrival ("Will be there in five minutes").

Perspectives on the individual activity patterns as well as our data reveal that the spatial component and a synchronization of it are essential in mobile conversations. The synchronization of the time dimension can also be tracked: The arranging over the phone allows for the conversation partners to track previously completed activities as well as – and this is probably crucial – a potential coordination of future activities. This behaviour is especially comprehensible when mobile communication is used in order to coordinate face-to-face meetings. Mobile phones allow for new forms of coordination between people. Ling and Yttri (1999, 2002), for instance, described micro-coordination, as an arrangement and re-arrangement of personal meetings and contacts via mobile phones. The observation of individual behavioural patterns and the way people synchronize them also sheds a light on this form of coordination. A case study from an observation episode at a lakeside illustrates that people even arrange for the termination of an activity or respectively the abandonment of the hereby intertwined place by mobile phone. In doing so one can lead over to the next activity: A woman phones her husband on leaving the bathing beach in order to notify him and to arrange for the meeting at home as well as for the beginning of the collectively agreed dinner. Frequently, mobile communication serves – in the course of activity patterns – to define these patterns and also to coordinate the course of these patterns of different people. Thus, a flexible planning of subsequent activities will be feasible. Coordination in this sense of the word has to be interpreted as a merger of various activities of two or more individuals. Hereby the mobile phone is used to press ahead, co-ordinate, and possibly synchronize activities. If one now sees individual activities like the bathing at the beach in the context of a person's course of activities (this person may have done the grocery shopping beforehand and goes home, in order to have dinner) than it is more comprehensively about co-ordination of activity patterns, and thus procedures which are, in this case, co-ordinated by husband and wife. The mobile phone allows people to plan their activities in all their timely and spatial procedures and thus also simplifies the happening of social events. A softening of time (Ling / Yttri 2002) goes along with this synchronization process. From this perspective one has to add in the discussion about social effects of mobile communication an argument for the facilitation of social cohesion and against the blunting of social competencies hypothesis (Geser 2005; Katz 2006; Ling 2008).

The co-ordination of activity patterns and the mobile phone

Regarding everyday life structures and daily routines one can speak of an index-linking through mobile communication. That is, people's activity patterns do not only boil down to physical meetings or to the availability at specific places (at home, in the workplace) as suggested by landline telephony (cf. Geser 2005). Mobile communication leads to the possibility of direct exertion of influence on the activities of others and thus to the possibility of changing and synchronization of patterns. Mobile phones have established an immediate and flexible communicating. An insight into activity patterns allows for an extended perspective on these associated social phenomena. This initiates an understanding of influences of time and space structures on everyday procedures of people and their relations to one another. This approach may be especially fruitful for an analysis of networks and their communicative structure and dynamic (cf. also Campbell / Ling 2009.)

Here, with regard to the analysis of interaction arrangement, a classification of anticipation strategies that mobile communicators implement with regard to communication partners has been made. These strategies – and here the linkage between both analysis threads becomes apparent – become obvious also from the perspective on activity histories. One case of our observation study shall serve as an example here: A fifty-year-old man is calling from a public phone[16] which is located at a central place in the city centre of the German city of Erfurt. In this telephone conversation he tries to agree with an acquaintance on a meeting for the next day. The follow-up interview clarifies that the anticipation of activity patterns and thus the resulting availabilities on (mobile) media is of great importance. Once the man explained the call and the reasons for it, like why he is making the call here and then, he also outlines his own plans, his location (city centre) appointments respectively his time windows (i.e. dentist appointment) for the day. He also outlines his friend's timetable which is known to him. He further factors things in like that he could not call his friend earlier as he would have been in the middle of a tennis match:

> "[...] because I have an appointment at the dentist and the next appointment is already pressing, when he is done with tennis then we need to arrange for a meeting and so on. Well, how it is like."

This form of co-ordination illustrates a conjunction of activity patterns of different people. The pursued strategy comprises of, firstly, an anticipation of activity patterns of the other respectively the others as well as, secondly, the choice of an appropriate situation for co-ordination which then again is transferred into an establishment of contact by means of mobile communication. Clearly, the knowledge about other people's everyday life requirements and structures has got a pivotal significance in this process. This knowledge is linked to the relationships of the people communicating. Thus, relationship specific behaviour is clearly revealed in these co-ordination and anticipation processes. This knowledge is factored into the considerations of a person while in preparation for a potential establishment of contact. Such an assigned situational status of communication partners, which results from an anticipation of both one's own as well as the real situational integration of the communication partner, explains the behaviour of communicators in public space. An example for this serves an episode of a young man who was observed waiting at Erfurt's Domplatz square during one of the observations. He has got a bicycle with him and lingers at the detached and by stairs surrounded monolith while talking on the phone (see picture 1). During the nearly fifteen-minute telephone conversation he lingers around his bike and the monolith (cf. the aspect of lingering,

16 During the observation studies about communication in public space several episodes of phone column usage were recorded. These kinds of phone columns are open versions of telephone booths and are more and more common in Germany. Hereby the mobile phone was often used in order to produce telephone numbers. Reasons for not using the mobile phone as means of communication were calls to landline phones or international calls as these are cheaper from roadside phones. This was also the case in our depicted example. Other explanations for using roadside phones were empty phone batteries or no credit left.

Höflich 2006), fiddles from time to time with his bicycle and is completely sunken into his telephone conversation. In the subsequent interview he explains the phone call: He was on his way to his mother's birthday party when all out of the sudden his girlfriend called him on the phone. She was just returning from a trip abroad and was contacting him the very first time after her arrival which is why he found it especially worthwhile to interrupt his ride in order to talk to her.

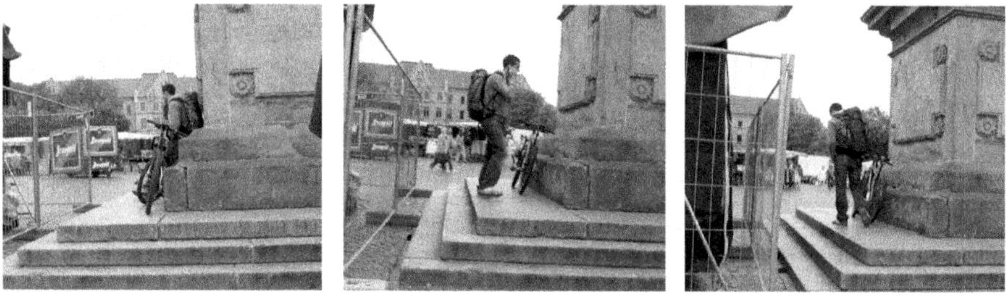

Illustration: observation episode in 2007 at the Domplatz in Erfurt

The case study exemplarily clarifies how important relationship characteristics as well as their current status are for people's communication behaviour in public space. This also conveys the dynamics of relationships. The behaviour of the young man would have been a different one if he had already met or talked to his girlfriend in the morning. Interaction scripts are established relationship specific and in doing so they are permanently updated, corresponding to the development of a relationship. Moreover, ways of conduct in mobile communication in public space are also established relationship specific. Concerning this matter, our study revealed that people reflect differentiated ways of conduct variously. While some participants reported that their communication behaviour differentiates (distance, volume etc.) depending on with whom they are communicating, others again reported that they generally maintain a constant behaviour.

Our findings show that the individual activity patterns are influenced by people's respective everyday life structures such as temporal specifications (working hours), spatial conditions (place of residence), and forms of mobility (itineraries) as well as regulations that go along with them. This yields specific patterns for the mobile communication behaviour such as the initiation or the reception of contact. The present findings reflect via these actions and conducts which are tied to communication to the fact that people who find themselves in these situations a priori align themselves independently of concrete communication, and only later on these people classify themselves into their current state of mobile availability or willingness to communicate. They define something like a "mobile communication mode". This became clear in some observations as modifications in the phone's settings were undertaken such as "silent mode" for no availability or only limited availability and ready for communication or "alarm" for available. Even the shutting off or the turning on of the mobile phone after leaving the workplace indicates as such an action of defining a individual communication situation. An internalization and anticipation of mobile communicating is outlined in another episode: In an interview a man who is typing on his phone while standing around at Erfurt's Anger square, expresses that, in expectation of a call, he is looking through the call list to make sure that he has not missed the call:

> "Aye, I am expecting a call. [...] In the list of missed calls was nothing which is why I put the phone back in my bag."

The motion in public space respectively the trespassing or the abandonment of public space corresponds to a systematic relevance (see Höflich / Kircher in this volume): the termination or the beginning of mobile media usage could often be observed on entering or exiting of buildings (e.g. post offices, museums or passages). However, leaping to conclusions is not possible here. Only a differentiated understanding of individual patterns of conduct allows for conclusions such as whether or not mobile communication has got priority over the changing real situations and thus whether a situation (whether public, semi-public or private) is abandoned or sought for. It is just as possible that mobile communication gives way to either the open situation or the closed one. The situational changes often go along with shifts in the 'communication mode', however, one cannot per se attribute a tendency of predominant mobile phone usage in public or in private space. This behaviour is so much more differentiated. It became apparent that concrete and for people significant situations are made up of much smaller aspects, and thus these aspects determine mobile patterns of conduct. In order to understand the behaviour of the individual, it is important that the individual's respective subjective perception of the communication process and the environment is captured. With some people this is implemented through a modification of their mobile phone settings. It remains open – and here it is necessary to find out the motives and reasons – why other people do not act in such a way. These patterns of conduct are closely intertwined with personal activity patterns and demonstrate that a lot of people deliberately define a "mental communication mode" for themselves.

The insight that subjective perceptions as well as evaluation are presumptions of communicative conduct goes along with the findings of our two analysis perspectives. The subjective component is both crucial for a sufficient understanding of individual activity patters as well as for a comprehensive understanding of situational interaction arrangements. An important link between the two perspectives results from these findings.

Conclusion and theoretical consequences

Differentiated interaction strategies with regard to mediatized communication in public space were identified within the frame of the study. The participants of media supported conversations have to anticipate the current situation of others and need to bring their actions in line with this construction although they do not directly experience these situations. The findings prove that communicators in public space need to bear in mind the requirements of the context since the specific regulations of the context as well as the fundamental standards of interpersonal co-habitation are of importance. It has proven that people evaluate some of the differently regulated places in public space as more appropriate for certain actions or conversations on mobile phones than others. Simultaneously, places obtain their specific characters only through the individuals' ways of usage. The analysis describes anticipation strategies which communicators implement with respect to their communication partners. This underscores the importance of social relationships for an understanding of the communication in public space. Anyhow, the concerned parties develop conclusions of conduct out of the anticipation of how they want to be treated in an actual situation by their medially or physically present counterpart. Such a conduct often presumes a fundamental knowledge about the everyday life structures of the callee.

This research describes furthermore ways of conduct managing attention during which a linked up observation of media-supported and face-to-face communication was enabled for. The findings have shown that the intrusion of a mobile phone user into the acoustic space of other present parties is a problematic aspect of mobile telephony. The individual evaluations of which matters can be or are allowed to be discussed in public are closely interwoven with this aspect. The assumption that adolescents, who were already stamped by mobile phones in their socialization, set these boundaries differently was dismissed by the empirical findings. The findings point out those repertoires of interaction forms were actually enhanced by an additional media component. Such an enhancement requires from the individuals the willingness to admit new situations into the repertoire of already existing situations and thus the possibility to respectively use the media's potential under the full adherence of the universal code of conduct.

Furthermore, the empirical analysis could contribute to a reflection of mediatized communication in relation to everyday life patterns of people by the use of comprehensive and through different methods generated data. Following this perspective, it was assessed that mobile communication in public space is linked to interruptions, bridging as well as the combination of people's activities. It became evident that mobile communicators in public space assess their actions with regard to temporal as well as spatial aspects. Mobile communication often serves with respect to activity patterns to assess their further course and to coordinate the course of patterns of various people. They use mobile phones in order to plan directly their actions with their temporal and spatial progress and thus ease social exchange and create a foundation for social events. These findings can be interpreted as an argument for the facilitation of social cohesion processes through extended possibilities of communication media. Even from the perspective of everyday life patterns, the already mentioned anticipation strategies could be identified. These strategies serve as a form of coordination and combination of various people's activities. It was compiled that the coordination and anticipation processes clearly convey relationship related behaviour. This knowledge is factored into the considerations of people who prepare for a potential establishment of contact. Such an assigned situational status of communication partners, which results from an anticipation of both one's own as well as the real integration of communication partners, explains the behaviour of communicators. In doing so, one establishes relationship specific behaviours of mobile communication in public space. The actual findings strongly refer to linkages between communication and behaviour. That is, that people assess the situation, in which they are in, independently from concrete communication and after that they define anticipatory their current status of mobile availability or communicational disposition – "mobile communication mode". The subjective perception as well as the evaluation based on situational classification or considerations with respect to everyday routines constitutes presuppositions of communicative behaviour.

Thus, on the basis of the empirical findings it is now feasible to graphically illustrate and extent the already described communication model (based on Höflich 2003 as well as 2005b) which bases on the contextual incorporation of media supported communication between callers and callees and considers the potential presence of third parties on both sides.

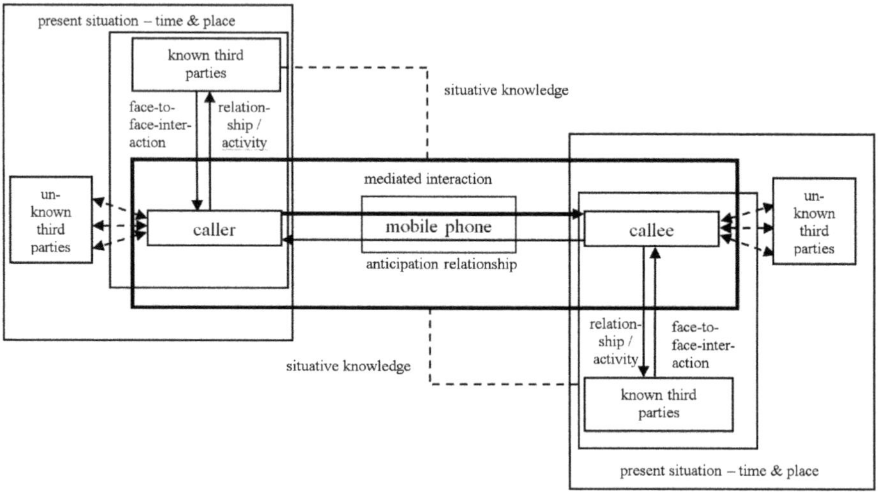

Illustration: integrative perspective of mediatized and direct communication.

Both the callers as well as the callees contextualise their own actions, spatial and temporal aspects hereby serve as landmarks. Furthermore, communication partners anticipate besides their own situation the situation of their medially placed counterpart. Based on this synchronization process they assess the possible progress of the forthcoming media situation. The situation anticipation process may happen on the caller's side before the phone call, the caller hereby draws on the knowledge of the callees daily routines. By indicating bilaterally the available capacities and action possibilities, a mutual agreement is concluded about the conversation. The relationships between the medially connected partners as well as the relationships between the present partners have got a significant impact on these negotiation processes. The fusion of these above mentioned aspects leads towards the observable arrangements. Thus, this illustrates that only after giving attention to all mentioned partial-phenomena of this complex negotiation process it is possible to capture media supported interactions, to classify them within the context of everyday life action and to make the happening mediatization describable on the basis of empirical data.

References

Angrosino, Michael (2007): Doing Ethnographic and Observational Research. Los Angeles [et al.]: Sage Publications.

Asendorpf, Jens / Banse, Rainer (2000): Psychologie der Beziehung. Bern et al.: Verlag Hans Huber.

Barbour, Rosaline (2007): Doing Focus Groups. Los Angeles [et al.]: Sage Publications.

Barker, Roger G. (1968): Ecological psychology. Concepts and Methods for Studying the Environment of Human Behaviour. Stanford/Calif.: University Press.

Bahrdt, Hans Paul (1996): Grundformen sozialer Situationen. Eine kleine Grammatik des Alltagslebens. München: C.H. Beck Verlag.

Banks, Marcus (2007): Using Visual Data in Qualitative Research. Los Angeles [u.a.]: Sage.

Berger, Peter L. / Luckmann, Thomas (2007): Die gesellschaftliche Konstruktion der Wirklichkeit. Frankfurt am Main: Fischer Taschenbuch Verlag.

Burkart, Günther (2000): Das Mobiltelefon und die Veränderungen der Kommunikation im sozialen Raum. In: Beck, Ulrich (ed.): Soziale Welt. Zeitschrift für sozialwissenschaftliche Forschung und Praxis. Volume 51, Number 2 (2000). Baden-Baden: Nomos Verlagsgesellschaft, 209-232.

Campbell, Scott W. / Ling, Rich (2009): Mobile Communication in Space and Time – Furthering the Theoretical Dialogue. In: Ling, Rich / Campbell, Scott W. (eds.): The Reconstruction of Space and Time. Mobile Communication Practices. New Brunswick and London: Transaction Publishers, 251-259.

Canary, Daniel J. / Dainton, Marianne (2003): Maintaining Relationships Through Communication. Relational, Contextual, and Cultural Variations. London et al.: Lawrence Erlbaum.

Chapin, Stuart F. (1974): Human Activity Patterns in the City. Things People Do in Time and Space. New York et al.: John Wiley & Sons.

Corbin, Juliet / Strauss, Anselm (1996): Grounded Theory. Grundlagen qualitativer Sozialforschung. Weinheim: Beltz.

Dourish, Paul (2004): What We Talk About When We Talk About Context. In: Personal and Ubiquitous Computing, Volume 8, Issue 1, 19 – 30.

Duck, Steve (2007): Human Relationships. Los Angeles et al: Sage.

Fortunati, Leopoldina (2005): Is Body-to-Body Communication Still the Prototype? In: The Information Society: An International Journal. Volume 21, Number 1, Philadelphia: Routledge, 53-61.

Fox, Kate (2001): Evolution, Alienation and Gossip. The Role of Mobile Telecommunications in the 21st Century, Social Issues Research Centre. URL: http://www.sirc.org/publik/gossip.shtml (February 26, 2009).

Garfinkel, Harold (1984): Studies in Ethnomethodology. Cambridge: Polity Press.

Gergen, Kenneth (2002): The Challenge of Absent Presence. In: Katz, James; Aakhus, Mark (eds.): Perpetual Contact: Mobile Communication, Private Talk and Pubic Performance. Cambridge: Cambridge University Press, 227-241.

Gehrau, Volker (2002): Die Beobachtung in der Kommunikationswissenschaft. Konstanz: UVK Verlagsgesellschaft mbH.

Geser, Hans (2005): Is the Cell Phone Undermining the Social Order? Understand Mobile Technology from a Sociological Perspective. In: Glotz, Peter / Bertschi, Stefan / Locke, Chris (eds.): Thumb culture. The Meaning of Mobile Phones for Society. Wiesbaden: Transcript Verlag, 143-157.

Glotz, Peter / Bertschi, Stefan / Locke, Chris (eds.): Thumb Culture. The Meaning of Mobile Phones for Society. Bielefeld: Transcript Verlag.

Goffman, Erving (1971): Verhalten in sozialen Situationen. Strukturen und Regeln der Interaktion im öffentlichen Raum. Gütersloh: Bertelsmann Verlag.

Goffman, Erving (1972): Relations in the Public: Micro-Studies of the Public Order. New York et al: Harper&Row.

Goffman, Erving (1978): Interaktionsrituale. Über Verhalten in direkter Kommunikation. Frankfurt am Main: Suhrkamp Verlag.

Goffman, Erving (2002): Wir alle spielen Theater. Die Selbstdarstellung im Alltag. München: Piper Verlag.

Höflich, Joachim R. (2001): Das Handy als "persönliches Medium": Zur Aneignung des Short Message Service (SMS) durch Jugendliche. In: Kommunikation@gesellschaft, Volume 2, Number 1. (see also: http://www.soz.uni-frankfurt.de/K.G/B1_2001_Hoeflich.pdf, 2.03.2009).

Höflich, Joachim R. (2003): Part of two Frames. Mobile Communication and the Situational Arrangement of Communicative Behaviour. In: Nyri, Kristóf (ed.): Mobile Democray. Essays on Society, Self and Politcs. Wien: Passagen Verlag, 107-116.

Höflich, Joachim R. (2005a): The Mobile Phone and the Dynamic between Private and Public Communication: Results of an International Exploratory Study. In. Glotz, Peter / Bertschi, Stefan / Locke, Chris (eds.): Thumb Culture. The Meaning of Mobile Phones for Society. Bielefeld: Transkript, 123-135.

Höflich, Joachim R. (2005b): An mehreren Orten zugleich: Mobile Kommunikation und soziale Arrangements. In: Höflich, Joachim R. / Gebhardt, Julian (eds.): Mobile Kommunikation. Perspektiven und Forschungsfelder. Berlin: Peter Lang Verlag, 19-41.

Höflich, Joachim R. (2005c): A Certain Sence of Place. Mobile Communication and Local Orientation. In: Nyíri, Kristóf (ed.): A Sense of Place. The Global and the Local in Mobile Communication. Vienna: Passagen, 159-168.

Höflich, Joachim R. (2006): Places of Life – Places of Communication: Observations of Mobile Phone Usage in Public Places. In: Höflich, Joachim R. / Hartmann, Maren (eds.): Mobile Communication in

Everyday Life: Ethnographic Views, Observations and Reflections. Berlin: Frank und Timme, 19-51.
Höflich, Joachim R. (2007): Zur Kommunikationskultur: Handy und SMS. In: Rosenstock, Roland / Schubert, Christiane / Beck, Klaus (eds.): Medien im Lebenslauf: demographischer Wandel und Mediennutzung. München: kopaed-Verlag, 139-161.
Höflich, Joachim R. (2008): Reinventing the old? New Communication Technologies and Changing Practices of Communication. In: The Role of New Technologies in Global Societies. Theoretical Reflections, Practical Concerns, and its Implications for China. Department of Applied Social Sciences, The Hong Kong Polytechnic University. Hong Kong 2008, 373-382.
Höflich, Joachim R. (2009): Mobile Phone and Emotional Stress. In: Fortunati, Leopoldina / Vincent, Jane (eds.): Electronic Emotions. The Mediation of Emotions via Information and Communication Technologies. New York et al.: Peter Lang, 63-83.
Höflich, Joachim R. / Hartmann, Maren (2007): Grenzverschiebungen: mobile Kommunikation im Spannungsfeld von öffentlichen und privaten Orten. In: Röser, Jutta (ed.): MedienAlltag: Domestizierungsprozesse alter und neue Medien. Wiesbaden: Verlag für Sozialwissenschaften, 211-221.
Höflich, Joachim R. / Linke, Christine (forthcoming): Mobile Communication in Intimate Relationships: Relationship Development and the Multiple Dialectics of Couples' Media Usage and Communication. In: Ling, Rich / Campbell, Scott (eds.): The Mobile Communication Research Series: Volume II, Mobile Communication: Bringing Us Together or Tearing Us Apart? Piscataway, New Jersey: Transaction books.
Katz, James E. (2006): Mobile Communication and the Transformation of Daily Life: The next Phase of Research on Mobiles. In: Knowledge, Technology, and Policy, 19 (1), 62-71.
Katz, James E. (2008): Handbook of mobile communication studies. Cambridge et al.: MIT Press.
Knapp, Mark L. / Vangelisti, Anita L. (2005): Interpersonal Communication and Human Relationships. Boston: Allyn & Bacon.
Klamt, Martin (2007): Verortete Normen. Öffentliche Räume, Normen, Kontrolle und Verhalten. Wiesbaden: VS Verlag.
Kowal, Sabine / O'Connell Daniel C. (2005): Zur Transkription von Gesprächen. In: Flick, Uwe / Kardorff, Ernst von / Steinke, Ines (eds.): Qualitative Forschung. Ein Handbuch. 4. Auflage. Reinbek bei Hamburg: Rowohlt Taschenbuch Verlag, 437-447.
Krotz, Friedrich (2001): Die Mediatisierung kommunikativen Handelns: Der Wandel von Alltag und sozialen Beziehungen, Kultur und Gesellschaft durch Medien. Wiesbaden: Westdeutscher Verlag.

Krotz, Friedrich (2007): Mediatisierung: Fallstudien zum Wandel von Kommunikation. Wiesbaden: Verlag für Sozialwissenschaften.

Kruse, Lenelis (1986): Drehbücher für Verhaltensschauplätze oder: Scripts für Settings. In: Kaminski, Gerhard (ed.): Ordnung und Variabilität im Alltagsgeschehen. Göttingen et al.: Hogrefe, 135-153.

Lamnek, Siegfried (1995): Qualitative Sozialforschung. Band 2: Methoden und Techniken. 3. korrigierte Auflage. Weinheim: Psychologie Verlags Union.

Lamnek, Siegfried (2005a): Gruppendiskussion. Theorie und Praxis. Weinheim [et al.]: Beltz PVU.

Lamnek, Siegfried (2005b): Qualitative Sozialforschung. Lehrbuch. 4. vollst. überarb. Auflage. Weinheim [et al.]: Beltz PVU.

Ling, Rich (2002): The social juxtaposition of mobile telephone conversations and public spaces. URL: http://www.richardling.com/papers/2002_juxtaposition_public_spaces.pdf (March 2nd, 2009).

Ling, Rich (2005): Das Mobiltelefon und die Störung des öffentlichen Raums. In: Höflich, Joachim R. / Gebhardt, Julian (eds.): Mobile Kommunikation. Perspektiven und Forschungsfelder. Berlin: Peter Lang Verlag, 115-133.

Ling, Rich (2008): New Tech, New Ties: How Mobile Communication Is Reshaping Social Cohesion. Cambridge, London: MIT Press.

Ling, Rich / Yttri, Brigitte (1999): Nobody Sits at Home and Waits for the Telephone to Ring: Micro and Hyper-coordination through the Use of the Mobile Phone. Report 30/99. Kjeller: Telenor Research and Development.

Ling, Rich / Yttri, Brigitte (2002): Hyper-coordination via mobile phones in Norway. In Katz, James / Aakhus, Mark (eds.): Perpetual contact: Mobile communication, private talk, public performance. Cambridge: University Press, 139-169.

Mey, Günter / Mruck, Katja (eds.) (2007): Grounded theory reader. Köln: Zentrum für Historische Sozialforschung.

Miller, Gerald R. / Steinberg, Mark (1975): Between People. A New Analysis of Interpersonal Communication. Chicago et al.: Science Research Associates Inc.

Morgan, David L. (1997): Focus groups as qualitative research. Thousand Oaks: Sage Publications.

Scherer, Klaus R. (1979): Kommunikation. In: Scherer, Klaus R. / Wallbott, Harald G. (eds.): Nonverbale Kommunikation. Forschungsberichte zum Interaktionsverhalten. Weinheim, Basel: Beltz Verlag, 14-24.

Sennett, Richard (1986): Verfall und Ende des öffentlichen Lebens. Die Tyrannei der Intimität. Frankfurt am Main: Fischer Verlag.

Weilenmann, Alexandra / Leuchovius, Peter (2004): "I'm waiting where we met last time": Exploring everyday positioning practices to inform design, in Proceedings of NordiCHI 2004, Helsinki.

Social aspects of mobile communication

Mobile media and affectivity: some thoughts about the notion of affective bandwidth

Amparo Lasén

Introduction

This chapter proposes a reflection on the complex relationship between emotions and technologically mediated communications and relationships, focusing on the concept of affective bandwidth. One of the consequences of the growing presence and use of mobile media is their participation in the emergence, shaping, display, management, control, expression and experience of positive and negative emotions, as well as in the social learning process of affective experience and communication. The ways we use and consume mobile media are redefining the realm of intimacy (Pertierra 2005; Prøitz 2005; Hjorth 2008; Tomita 2006; Crawford 2009). Mobile media users enjoy an emotional relationship with their devices and applications and are emotionally attached to them (Lasén 2004; Vincent 2005). This is partly due to the intrinsic emotional character of human communication, and also because mobile media have become technologies of intimacy (Haraway 1991; Fortunati 2002) and technologies of the self (Foucault 1988), which are present in most of the settings of everyday life and are part of contemporary embodiment processes.

The content of this chapter is grounded on empirical research, though it is not going to be discussed here (Lasén 2005; Lasén / Gómez 2009; Lasén 2009). Mobile phones, which are increasingly becoming mobile media (Hjorth 2008; Goggin / Hjorth 2009), are at the centre of this research, in convergence with other devices and applications. This empirical background relates firstly to the implications for contemporary embodiment processes and the redefinition of intimacy of digital photography, regarding self-portraits, their exchange in the Net or through mobile phone applications (MMS, Bluetooth), as well as their growing presence online, from Social Network and Sharing sites, to Online Dating sites or Instant Messaging exchanges (Lasén / Gómez 2009). Secondly, research about mobile phones and couples relationships[1] focusing on communication,

1 "Nuevas tecnologías de la comunicación y rearticulación de las relaciones de género: emergencia, expresión y gestión de los conflictos en la pareja" Research Project fundend by the Spanish Ministery of Science and Innovation, carried out with my colleagues of the University Complutense of Madrid, Elena Casado, Rubén Blanco, Concepción Gómez and Antonio García (2009-2011). 'La contribución de los usos del teléfono móvil a la configuración y transformación de redes, vínculos y subjetividades' by the Complutense

personalisation and conflicts (Lasén 2009) provides material and examples for this discussion. Thirdly, previous research about mobile phones and emotions, part of an European cross-national study about mobile phone users and uses, helped to ground a reflection about what an affective mobile phone could be (Lasén 2005, 2004).

Emotional turn

Since the mid-eighties of the last century, an increasing interest in emotions and affectivity has characterised social sciences and social practices. This affective turn (Clough / O'Malley 2007) is related to social changes in Western societies in the way of considering emotions, subjectivity and their relationship to rationality. It comes after several decades of critique and questioning of the traditional rationalist view of emotions. This conception considers emotions, at best as some luxury belonging to art and idleness and, at worst, as an obstacle to clear judgement. Therefore emotions had been relegated to private space and intimacy. As Norbert Elias (1982) points out discipline and the self-constraint of emotions and urges became necessary in societies with growing complexity and social differentiation. In these rationalistic societies, 'emotional' is not only an adjective simply denoting that something or someone is related to emotion, but also it becomes a pejorative adjective. Those characterised as emotional – women, children, primitives – were thought to lack the fundamental ability to think and decide rationally, and were placed in the lower levels of the social and even human hierarchies[2].

The social changes in Western societies that have modified this picture have been analysed under the label of the crisis of modernity, postmodernity, reflexive modernity or post-industrial societies. In contemporary society the expression of one's subjectivity becomes almost a claimed right.

Research Group Cibersomosaguas (2007-2008); Proyecto Complutense 2006 "La mediación de subjetividades e identidades sociales a través de la telefonía móvil", with my colleagues from the University Complutense of Madrid, Angel Gordo and Lucila Finkel.

2 The opening paragraph of the preface of *Affective Computing* by MIT researcher Rosalind Picard clearly summarizes, expressed with her personal unease, the dominant Western conception of emotions:
"I never expected to write a book addressing emotions. My education has been dominated by science and engineering, and based on axioms, laws, equations, rational thinking, and a pride that shuns the 'touchy-feely.' Being a woman in a field containing mostly men has provided extra incentive to cast off the stereotype of 'emotional female' in favor of the logical behavior of a scholar. For most of my life my thinking of emotions could have been summarized as: "Emotions are fine for art', entertainment and certain social interactions, but keep them out of science and computing" (Picard 1997: X).

> "Our entertainment, our social values, even the work we do and the governments we elect are all to be understood in terms of satisfaction, pleasure, like and dislike, excitement and boredom, love and hate. A world where we once knew ourselves in terms of values and identities has given way to the uninterrupted intensities of elation and grief, triumph and trauma, loss and achievement; birth, death, survival, crime, consumption, career are all now pretexts for emotion" (Mansfield 2000: 2).

For more than thirty years now, social scientists have analysed these changes in the social considerations of emotions and feelings, either to denounce the tyranny of intimacy and the vanishing of public space (Sennett 1977) or to criticise the excess of rationalism, acknowledging that emotions are not only individual and psychological features but fundamental aspects of social life (Maffesoli 1993). Under the term 'orgiasm', Maffesoli includes the orgiastic, emotional, sensual and sexual aspects of social life, which constitute a constant factor of sociality. He follows Durkheim's view that emotions are the grounds of the making of society, the "glue" of social solidarity and social cohesion. Emotional connectedness is the ultimate basis of social bonds. For Durkheim, collective effervescence or collective excitement, a state of heightened emotion, experienced in rituals, festivals and celebrations, produced by the sharing of strong feelings and emotions among the participants, is necessary to sustain and create social bonds, and is therefore necessary to the maintaining of any community and society. Rituals magnify and intensify any shared emotion in the participants, eliciting a feeling of solidarity. This solidarity arises from different kinds of shared feelings: from the joy of celebrations to the sadness of grief and mourning (Durkheim 1915). Thus the tension between a civilization process marked by the restraint, shield and control of emotions and bodies, and the necessary emotional atunement of social life would be one of the features of modernity and its paradoxes.

Besides the different diagnostics and evaluations, what has been manifested in the realm of social sciences these last decades is the importance of seizing the role of emotions in the constitution of social experience, in order to understand what is at stake for people in everyday life (Lutz / White 1986: 431). This is, the recognition that emotions are necessary to understand social experiences, as they are part of these experiences and are also shaped by them. Therefore emotions are subject to cultural, collective and historical variations. Feelings are socialised (Hochschild 1979). This socialisation entails learning how to feel, display and interpret emotions in different affective cultures. Emotional labour and management characterise a big part of the way people experience affectivity, and both are increasingly mediated by ICTs.

This emotional turn can be traced in the development of a sociology and anthropology of emotions (Rosaldo 1984; Lutz / White 1986; Franks / McCarthy 1989; Kemper 1991; Le Breton 1998; Scherer 2001), often grounded in a renewed interest and research about affects in neurology (Damasio 1994). This attention to emotions arises as part of the effort to counteract the disembodied character of classic social theory, strongly supported and initiated by feminist theorists and women studies, and therefore goes with the centrality of subjectivity and embodiment in socio-cultural studies (Turner 1984; Featherstone et al. 1991; Mansfield 2000). This turn happened as well in culture and media studies with some authors' claim to shift the centrality from meaning to affect (Grossberg 1992) acknowledging that culture and also media cannot be totally accounted for just as "signifying practices".

Computing and technological design and usability are also part of this turn, with the emergence and development of affective computing, emotional usability and emotional design. Affective computing refers to giving computers the ability to recognise, express and have emotions in order to improve human-computer interaction (Picard 1997, 2003; Picard et al. 2004), as people tend to interact socially and emotionally with computers (Reeves / Nass 1996). Applications of affective computing are found in different shapes (wearables, autonomous agents, robots, software) and in different areas, such as entertainment, learning systems, simulations, surveillance and security applications (cf. http://affect.media.mit.edu/).

Design and usability research have developed emotional awareness as well, taking into account pleasure, enjoyment and fun (Norman 2003). Designers are concerned with designing context for experiences for different users with different emotional styles, considering the presence of the objects in the users' life, their different expressions and not only the use. Devices not only have functions but also expressions that derive from their presence in the users' everyday life. This is a new way of understanding design that adds, to the traditional engineering and scientific skills, the support of psychology, sociology and arts. The aim of both affective computing and emotional design is to make ICTs uses and experiences an emotionally richer interaction, multi-media and multi-sensuous, which would improve communication and task performance. They wish to contribute to people's awareness of their emotional style, and, in the case of affective computing, to make computer devices a kind of emotional intelligence tools.

Nowadays there is an interest in emotions by the public at large, manifested for instance in media and media studies, marketing, advertising, politics and political communication, or in the notion of emotional intelligence (Gardner 1973; Goleman 1997) being embraced by educators and businessmen alike. This emotional turn seems to be another example of the interaction between forms of academic and scientific knowledge on the one hand, and other social discourses and practices on the other. This growing presence and interest about affectivity is also related to the renegotiation of the boundaries between private and public, leisure and work, personal and professional. Emotions which belonged to privacy, leisure and personal realms are now mobilised in public and working activities, and considered to be ways of augmenting productivity and efficacy. A phenomenon well seized in the term "playbour" used for instance to describe free labour, fan labour and peer production, as in the digital games industry (Küblich 2005). In these uses and practices emotions fall under the reach of technological rationality, aimed to rise productivity, to bring more efficacy and efficiency to working tasks, to make time use more productive too.

This rationality is also found in the way people use technological devices for interpersonal and affective communication and relationships. For instance, in our research about couples and mobile phones, housewives described their use of the mobile for family matters and exchanges using the same terms "efficacy", "efficiency", time spare, that their husbands used to describe the advantages of the mobile phone use for work. Another example can be found in online dating sites advertising, which puts forward the advantage of their services in these terms of productivity, efficacy and time. These considerations are found as well in the accounts of users to describe their motivations and practices in their online search for sexual or loving partners. Regarding the renegotiation of boundaries between work and leisure, or professional and personal relationships, related to these aspects it is noticeable the mutual borrowing of social practices and Internet applications, such as Social Network sites where the same design architecture serves both. Another example can be found in "Speed Dating" gatherings, a direct application of economic rationality to the realm of matchmaking and dating. Shortly afterwards of their first apparition in the late 90's in the US and the UK, business speed dating were organised as well, as a way of providing contacts for professional working in the same branch. Moreover, similar mechanisms are proposed to improve learning activities (Maidment / Crisp 2007), revealing the growing fluidity between those different realms in contemporary societies, the articulation between affectivity, group dynamics, efficacy and learning.

The use of the word 'passion' for advertising and internal communication by the mobile phone operator Vodafone UK as in the slogan "passion for our customers" is an example of this positive attitude to emotions, being far from the rationalist view of passions as dangerous ways of blinding the mind and disturbing clear judgement. Another example of this positive presence of emotions in corporate practices and discourses is given by the Spanish mobile phone operator Movistar, which has called its internet application for mobiles: *e-mocion*.

The practices and spaces where emotions are expressed are no longer considered secondary, revealing the growing importance of leisure activities and popular culture, not only economically but also in terms of personal self-fulfilment. The findings of my previous research about young adults' temporalities in Paris and Madrid, (Lasén 2001), show that the expectations of young adults about work and their notion of what constitutes an ideal job (self-fulfilment, creativity, enjoyment, friendly relationships) come directly from the kind of practices and relationships characteristic of their leisure time. These individual views correspond to changes in the capitalist system and the modes of working organisation in enterprises. Increasingly they are concerned with the creation, realization and mobilisation of desires, hopes, beliefs, expressions of feeling, changes of sensibility, forms of intuitive intelligence and new communicative resources (Thrift 2006; Lazzarato 2004). Affect becomes an economic factor necessary to the governing of productivity, not only because affectivity is increasingly commodified and mobilised in the workplace, as affective labour and emotional labour theories state it (Hochschild 1983 and authors such as Hardt and Negri, Lazzarato, Virno, see Hardt's essay "Affective Labor" in www.vinculo-a.net/english_site/text_hardt.html), but also because economic value is measured affectively as well. The modes of evaluation and measurement necessary to capitalist productivity depend on the intelligibility that science brings forth, contributing to the shaping of what is measured[3] (Clough et al. 2007).

> "We consider this to be a shift away from a State project to temper, direct, and regulate the economy through consciously calculated intentionality (as in Keynesian economics) and toward a radical neoliberal governance of economy where the value of productive activity is no longer found in conscious and calculated intention, but rather in the play of uncertainty and the direct manipulation of affectivity. Whereas a post-World War II economy was subordinated to the calculations, goals, and intentions of the State, a radical neoliberal governmentality now subordinates its activities to the logic of a market economy and a rationality of affectivity. Here, the value of affect emerges adjacent to the production of use values for exchange and where the distinction between labouring and activity can no longer be maintained." (Ibid.: 74)

In this sense, Affective Computing software and applications would help to augment people efficacy and productivity, as well as increase surveillance by gathering knowledge and data about emotional styles and people's affects. Biometrics-based surveillance systems are just one of these examples. Such questions about privacy, surveillance and the disclosure of personal information, are not often addressed by Affective Computing research, which only recently has started to consider the ethical implications of its applications. However they address the ethical issue by proposing a contract between users and designers, as if they were the only actors involved in the uses and practices of such pieces of software (Reynolds / Picard 2004). Although an effort is being made in this field, mostly in the work of Reynolds, surveillance, harmful and unethical uses of affective computing are mainly thought as the consequences of misuse, accidents or some unruly individual behaviour (Reynolds / Picard 2005). The

3 The existence of this power-knowledge link between science, capitalist economy and neoliberal forms of governmentality does not mean that all form of science have to play that role (Haraway, Thrift).

consideration of whether it is right to contribute to affective labour and the mobilisation and surveillance of emotions for increasing work efficacy and productivity, is absent of such writings on ethics. This absence is probably related to the lack of socio-cultural considerations in their way of approaching emotions.

The study of emotions needs interdisciplinary approaches, from psychology, biology, neurosciences, sociology, anthropology, and philosophy, acknowledging the interplay between psychobiological and socio-cultural determinants that do justice to the complex, multi-component phenomenon called emotion. The first consequence of this complexity and multi-dimensionality is the lack of a precise definition and of a common agreement of what emotions are. Another obstacle to a unified definition is the gap between emotion as an abstract category, and the experience of emotion, which always concerns particular feelings (love, anger, resentment, shame), with distinctive features.

Being affected, being moved, being acted upon

Mobility is a quality shared by both emotions and mobile media. The original sense and etymology of emotions refer to agitated motion, mental agitation or feelings of mental agitation. Experiencing emotions is to be moved. This movement has got different dimensions: physical as movements experienced by the body and their organs; movements as changes in feelings and thoughts; and movement as the emergence of something that was not there, as the actualisation of something that was virtual, the movement that brings to existence something that was just potential. This mobility associated with affects and emotions entails the passage from one experiential state of the body to another. Affects are transitions (Massumi 2002: 25). Social versions of emotions consider them as "temporary transformative social roles", "as moves that adjust relations in ways that produce and convey evaluations and interpretations", interpretations in which people generate what they interpret as well (Myers 2007).

According to Spinoza this movement results in the augmentation or diminution in the body's capacity to act. Passions were conceived as 'ways of being acted upon', of being affected, of being moved, by other beings, objects, events and situations. The ability to affect and to be affected is intrinsic to the category of emotion. This ability is intrinsic to the body as well:

> "to have a body is to learn to be affected, meaning 'effectuated', moved, put into motion by other entities, humans or nonhumans. If you are not engaged in this learning you become insensitive, dumb, you drop dead" (Latour 2004: 205).

Positive emotions, called by Spinoza 'affects of joy', augment the power of acting, the range of things that we are capable to achieve, the abilities of our body. They derive from a good encounter with a person, an idea, an object or a situation, from being affected; that is, acted upon, in a good way. These affects act as a springboard, allowing us to overcome difficulties and to think more clearly. But emotions can also be negative, passions of sadness, resulting from bad encounters, which diminish our ability to act and achieve. In both cases, affects are emergent phenomena, bringing with them something that was not there before, which does not have a mechanical causal relationship to the conditions out of which they came (McCarthy 1989). The conditions never determine completely what will happen. Affects occur unexpectedly, subject to random encounters, and therefore are not under control of the person who experiences them. Nowadays mobile media act upon us through their affordances and constrains. They mediate, facilitate and help to shape our encounters with other people, institutions, images and ideas, which move and act upon us. Therefore mobile media can become themselves affected and affecting bodies.

Versions of emotions and dualism

When studying emotions we come across a plurality of terms: affect, emotion, passion, sentiment, feeling, and multiple definitions and conceptions. There are different versions of what emotions are and entail, corresponding to a coexistent diversity of definitions, controversies, knowledge and disciplines (Despret 2004). These versions reproduce diverse dualisms regarding the way of conceiving emotions such as, reason / emotion, body / mind, universal-natural / socio-cultural, internal / external, active / passive-reactive. Thus, the traditional Western approach defines emotions as opposed to reason. They are considered to be passive reactions to external causes, which elicit them. Situated in the body, they are therefore revealed and measurable in biological processes. Emotions and their bodily manifestations are considered to be universal, common to all human beings regardless of socio-cultural and historical variations. This version has helped to develop many techniques and modalities used to detect physiological signals of affects, such as facial expression recognition, speech prosody recognition, and pressure sensors. It is one of the main conceptual backgrounds of affective computing as well.

Following the influential work of Charles Darwin (1872), the tenants of this version consider that emotions are universal and adaptive: a biological navigation software developed through evolution to be used by individuals or groups. They would offer guidance for coping with the various problematic situations people typically encounter in their lives, as setting a goal in a situation-dependent way i.e. fear, whose major function would be to automatically set the goal of reaching safety. In *The expression of the emotions in man and animals*, Darwin splits emotions from individual experience, making them a series of absolute states that take possession of persons. Darwin paves the way to a "botanic of emotions" (Le Breton 1998: 39), where any emotion could be described and classified without reference to the people who experience it.

Several programs of cross-cultural research on emotion draw on Darwinian insights, such as the 'neurocultural' research programme led by Ekman (Ekman et al. 1992), which studies facial expressions of emotions. In this approach culture has a secondary role. For instance, cultural rules about how to display emotions are thought to interfere with emotional responses dictated by the innate affect program. Ekman and Friesen present a systematic study of the facial muscles involved in the expression of emotions called FACS (Facial Action Coding System). This is a laboratory exercise, designed to produce an objective truth about emotions. French anthropologist David Le Breton critiques this approach for being indifferent to real people engaged in affective relationships, and also because by focusing on the facial muscles, it overlooks the fact that emotions are embodied and performed in the voice and other body movements besides the face. Gestures are isolated and not taken into consideration as part of a performance in a social interaction. That is, the meaning of these different gestures and muscle movements resides in the culturally contextualised sequence of expressions, in a necessary relationship with the whole of other gestures, positions, movements, and intonations. Emotions analysed by naturalist researchers are considered primary and universal. But, paradoxically, they do not agree on which of these are primary emotions. In everyday life, emotions are not substances, fixed states found in the same form for the whole of the human species, but affective tonalities (Le Breton 1998: 50). They colour human behaviour and are ambivalent, expressed in mixed feelings. There is not a unique expression of emotion but countless shades of body and face, which account for the affectivity of a social actor in a given context. In everyday life we do not find people expressing joy, but joyful people, with their particular style, ambivalence and singularity. Darwinian approaches encounter the pitfall of the duality individual / emotion and the ambiguity of the concept of 'expression'. Research carried out in laboratories, using pictures and films, or consisting in the stimulation of a group of volunteers' facial muscles sustains a version of emotions that does not account for historical changes, social bonds and social relationships in which emotions are embedded.

When this naturalistic approach is doubled with a negative consideration of emotions as an obstacle to clear reasoning, a certain number of paradoxes arise, for instance regarding the role of emotions as a sign of the authentic self and state of mind. Considering emotions reactive, outside people's conscious control, would be the ground for this authenticity. Individuals would be betrayed by their body reactions, from gestures to blood pressure, which would reveal their true state. This view does not account for all the forms of emotional management and control, and the ways of expressing and displaying emotions, which are socially shaped and learnt. Then, the belief that passions can drive yourself away from your sense, that you are not yourself when too moved, as you are not able to see and reason clearly, forms a paradox with the supposed authenticity of emotions. Thus, we would be outside ourselves when we are most intimately connected with ourselves and our own vitality (Massumi 2002). This ambiguity is stressed by contemporary conceptions of emotions as they are popularised by advertising, corporate management strategy or commercial media, with that double rhetoric about the importance of being passionate and involved, by also of being in control of what happens, of one's life, relationships, activities and feelings. In these brief accounts of contemporary subjectivities that are the self-presentation profiles of online dating sites, this double character of being passionate and spontaneous, but in control as well, is found very often. This paradox about self-control and authenticity also points to the ambiguous status of intentional behaviour, are we more authentic, closer to our true self when not acting intentionally, then?

To the list of dualities concerning emotions quoted above, we could add another found in recent versions linked to Spinoza's definition, which opposes emotion and affect (Deleuze / Guattari 1987; Massumi 2002: 28). Emotions are considered to be individual, experienced as personal, a subjective content, fixed by social convention and language, endowed with meaning and function, subject to socialisation practices and strategies of emotional labour and management, of intentional control and performance. Whereas affect would be pre-individual, non intentional, unqualified, unrecognizable and, unlike emotions, would entail a "suspension of action reaction circuits and linear temporality" (Ibid.)[4]. Affect "denotes a more-or-less organised experience, an experience probably with empowering or disempowering consequences, registered at the level of the physical body, and not necessarily to be understood in linguistic terms" (Gilbert 2004). This duality accounts for the non-social and non-conscious aspects of affectivity, without renewing the old conception of considering emotions universal, natural, and authentic, but sharing somehow some aspects with that approach, (Shouse 2005) regarding the non-conscious and bodily character of affects. As the bodily "innate" movements involving facial muscles, viscera, the respiratory system, skin, blood flow changes and vocalisations, triggered by uncontrolled factors are considered to be the sign of affects. Affectivity is a form of 'visceral perception' preceding perception, accounting for the autonomy of affect regarding consciousness and language (Massumi 2002: 25).

4 Bearing in mind that subjectivities and subjective content, are not only related to the individual subject and social conventions, as complex dynamics, resistances and a multiplicity of actors are involved in the always dynamic subjectivation processes.

The study of emotions is narrowly linked to the analysis and study of subjectivities, which are also burdened by a similar dualism, or by the attempts to overcome it by proposing a monist solution, a synthesis or harmonisation of the opposed terms. Translating to the research about emotions what Elizabeth Grosz proposes for the study of subjectivities (Grosz 1994), we could see that the problem with dualism is not that it differentiates and particularises entities and concepts, but that these are placed in a particular relationship of opposition, hierarchy or exclusion, that is supposed to always exist, in any kind of situation or case. Following Grosz, instead of replacing dualism by monism, and therefore losing the particularities and differences of the terms forming the duality, a more productive approach would be to consider the articulations and disarticulations, resonances and dissonances between those different terms in each particular situation. How the body and the mind, the internal and the external, the collective and the individual, the biological and the cultural, as well as the different mediations and material conditions of our environment are mobilised in the emergence, shaping, expression, display, performance, experience and understanding of emotions. While being attentive to where these aspects are mobilised and who are the actors (people and objects) involved. This is to say, how the different approaches, the different versions of emotions, with their particular forms of knowledge, methodologies and research instruments, contribute to their shaping as well. As in the example of the naturalistic approach described above, the mechanisms of the laboratory experiments contribute to shape the emotions they measure and detect. This does not imply that all versions are equally valid. Following Despret's analysis, these different versions can be evaluated by their ability to recognise, resonate and dissonate with the other, and by the opportunities their methods and concepts create, allowing emotions to reveal themselves and to be considered in all their richness and complexity.

Then, we could recognise that the duality between affect and emotion quoted above can be considered differently as well, looking at how in everyday situations these individual and pre-individual aspects are articulated, for instance by observing media practices where affectivity emerges and is mobilised. Søren Mørk Petersen's research about the sharing of personal pictures on the Web offers a good example of this, revealing how in Flickr users' practices both forms coexist. The affectivity involved, the emotions shared and the quality and intensity of the sharing experience, related to the display in the pictures of the banality of everyday life, go beyond the function and meaning of the pictures shared, and relate to collective and common experiences. The subjective but not necessarily conscious, individual and intentional aspect of such practices is revealed by the difficulties of the amateur photographers to explain why they take and share those pictures (Mørk Petersen 2009). In our research about mobile phone uses and practices both dimensions of affectivity are found, sometimes entangled in the same situation: the emergence of unforeseen and unintentional emotional experiences, as well as the collective conventional shaping of affects, strategies of control, monitoring, accountability and modes of subjection (Lasén 2009; Qiu 2007; Green 2001).

Changes concerning the relation between affect and cognition in contemporary versions about emotions are relevant for our topic as well. After centuries of a Western tradition that considered emotions as opposed to reason, as an obstacle to clear thinking, contemporary versions from social sciences to neurology consider that emotions have an essential role in basic rational and intelligent behaviour. They are a necessary part of processes of evaluation, interpretation, expression, sensemaking, and regulation of exchanges. Emotion is a motivating and a guiding force in perception and attention and therefore strongly affects memory and the way we learn, prioritise, make choices and plans. Mobile media are involved in and transform all these activities as well.

When we make a decision, we evaluate alternatives. If we are unable to feel emotions, we are unable to place values on the different alternatives. If we cannot place values, then, there is no difference between the alternatives, and decision-making becomes seriously flawed. Hence without emotions, rational decision-making may not be possible. This important role of emotions in the process of reasoning has been studied by neurological researches on brain-damaged patients (Damasio 1994). In spite of the doubts that the current method in medical research of deriving general conclusions from the behaviour of impaired individuals could rise in other research fields. Too little emotion impairs decision-making, states Damasio, when describing how patients with brain damage, namely frontal-lobe disorders, appear to have normal intelligence but make disastrous decisions. Their brains are missing "somatic markers" that associate positive or negative feelings with certain decisions. Therefore, the articulation between affects and cognitions is also one of the aspects of the articulation between body and mind: as in the notion of "somatic markers" or in the consideration that emotions are thoughts felt in the movements of our bodies and their organs, "*embodied* thoughts, thoughts seeped with the apprehension that 'I am involved'" (Rosaldo 1984: 143). This physicality of thoughts is also highlighted by Peter Sloterdijk's concept of *Dyonisian Materialism* (Sloterdijk 1989). This notion calls for the recognition of body and performance as the basis of consciousness and subjectivities, highlighting the material and embodied conditions of consciousness and subjectivity.

These embodied feelings help to prioritise, to limit a mental search by nudging the person away from considering the possibilities with bad associations. Emotions are evaluations of circumstances that provide information about relations to other, objects and events. They are judgements about the world. Therefore, they are endowed with an ethical dimension, as it was made explicit in the title of Spinoza's treatise. This could correspond to what Michel Maffesoli defines as an ethics of aesthetics, reminding of the etymology of aesthetics as "feeling together", experienced with others.

Emotions present us with a non-verbal preconscious assessment of our current situation. Examples of the effect on perception are the fear-induced phenomenon of "tunnel vision" or the joy-induced state of "looking through rose-tinted spectacles". This evaluation is infused with a sense of expectancy. It is a preparation for action. Emotions provide the apprehension of a possible future through an interpretation of the past experience of the subject. This is one of their specific contributions to action and agency. Without emotions individuals would be lost in time, between a remote past and an inaccessible future (Kemper 1978: 185). However, as Deleuze and Massumi point out regarding what they call 'affects', the temporality mobilised is not a linear one, nor the result of an action-reaction mechanism.

Affective Computing adopts a naturalistic version of emotions, which recognises their link with cognition but ignores social issues and social sciences literature concerning emotions and their cultural variations (Despret 2004). Emotions are considered to be natural, universal, reactive, situated and measured biologically in the body and its organs. This choice is understandable because the simplified and systematic conceptualisation of emotions and emotional expression makes easier the formulation of emotional models susceptible to be implemented in software, robots and other computational devices. However this is an obstacle to the main objective of affective computing: the achievement of the emotional interaction between users and devices. This ignorance of the socio-cultural aspects of emotions is doubled with a lack of consideration of the dark side of emotions. Affective computing authors tend to identify emotion with positive emotions and forget that all affects are not affect of joy. Mobile media uses and practices entail many examples of these bad encounters that trigger passions of sadness, borrowing Spinoza's terms: harm, flaming, cyberbullying. Anger, anguish, anxiety, embarrassment, guilt, sadness are emotions as well, which are present, shaped and elicited in mobile media practices. Negative emotions also relate to difficulties to communicate. In all these cases a richer emotional experience would not augment people's abilities. As Spinoza points these affects diminish people's abilities to act and achieve.

Following the scientific and biological categorisation of emotions, these researchers do not seem to acknowledge the ambivalence and mixed feelings involved in any emotional experience, and that affective technologies or emotionally richer human-computer interactions would not eliminate the frustration and paradoxes inherent to social bonds and human communication. For instance, mobile phones can increase the possibility of making good encounters, of being connected to those who bring us joy and relieve our sadness and stress. But being a device of social connectivity, they also create the occasion for bad encounters (unwanted calls, excessive requests, disturbing interruptions) and are often the witness to the lack of contact, when no call or message is received. They also mediate the expression of negative passions, such as anger, hate, anguish and anxieties.

Affective bandwidth

The broad categories of mobile media and social media that are relevant for our topic cover a multiplicity of devices (mobile phones, computers, laptops, PDAs, digital cameras, webcams, mp3 players) with an ever growing number of applications, such as email, SMS, Instant Messaging (IM), and different forms of social media, such as Internet fora, social network sites (SNS), podcasts, sharing sites, blogs, photoblogs and microblogs. This multiplicity is also multimedia (images, sounds, texts) and multi-sensuous (view, hearing, voice and touch) (cf. mobile phones as multi-sensuous devices, Vincent 2006). Bearing in mind the particularities derived from this multiplicity and the implications of the diverse forms of technological and social convergence[5], this section explores the notion of affective bandwidth, which is used to understand how mobile media are contributing to affectivity.

This term refers to the amount of affective information relayed through a particular device or application (Picard 1997: 57). It is currently used in affective computing, software and videogames design. The affective bandwidth of a channel corresponds to how much affective information the channel lets through. Information is conceived as "discrete units or states internal to an individual that can be transmitted in a loss-free manner from people to computational systems and back" (Boehner et al. 2007: 275). But this is a particular kind of information that not only concerns the kind of feeling or emotion but also the shades, degrees, the emotional tonality of the exchange and the affective state of the participants. Affective computing researchers adopt a version of emotions in which they are not opposed to cognition, on the contrary they are part of it, but in doing that they reduce emotion to cognition standards, failing to account for the particularities of emotions and their socio-cultural implications (Ibid.).

---

5 I understand by social convergence not only "when disparate social contexts are collapsed into one" (Boyd, 2008) but also the flow of different interactions and social functions embodied in the same device or application, such as connectivity, coordination, everyday organisation, control and monitoring, entertaining, affective communication, emotion management, accessibility, heteronomy and autonomy… All these aspects are made visible and physical, through their inscription in the device as sounds, images, numbers and texts. Both aspects of social convergence make difficult the control of uses and exchanges and can elicit conflicts and dissonances.

From affective computing researchers' studies, it can be deduced that this bandwidth increases when more senses are involved and also when moving from diachronic to synchronic communication. Then, affective information would be shaped by immediacy and delay. According to this approach, email and SMS would have the lower affective bandwidth, followed by IM, then phone conversation, then video conference and the maximum would be in face-to-face exchanges. It is interesting to note how this sequence resembles to the communication stages of people meeting in online dating sites: first emails and IM, then webcam and / or phone (and when phone numbers are exchanged also SMS) and finally the face-to-face date. However the intensity of the feelings and emotions experienced in technologically mediated communication does not necessarily follow that path. "Colder", asynchronous, written media do not always facilitate colder emotional experiences. From cybersex to the anger or love bursts caused when writing or reading particular emails and SMS, many are the everyday examples of intense emotional experiences afforded by this media. In the example of cybersex, talk and written chats "function as sex" (Atwood 2009). They are forms of sexual intercourse. Regarding the senses and the body, one should not forget that in spite of lacking of clues about the physical presence of the other in many forms of technologically mediated communication, bodies are there, involved and affected. The participants feel in their bodies and their senses the effects of the affective experiences they are living.

It is usually assumed that technology-mediated communication always has less affective bandwidth than person-to-person communication. This comparison between face-to-face and technology-mediated communication does not take into consideration the specific emotional affordances of ICT. It is not only a question of bandwidth, but of shades, of emotional tonality. For instance asynchronous forms of communication, from love letters to e-mails and texts, provide a way of displaying the emotion of the sender and of eliciting emotional responses in the receiver, which is different from the face-to-face communication, without being necessarily less intense or inferior. In many occasions the absence of the face-work and self-control required by synchronous and co-present forms of communication facilitates the emergence of affective occasions. In these cases a particular form of emotion, embarrassment, is lowered, making easier the emergence of other kind of affects and affective information.

Mobile phones, namely through texting, allow the avoidance of some of the risks, dangers and embarrassing social consequences of certain kind of emotionally charged exchanges. In this case the lower threshold for making contact through texting, compared to conversation, facilitates emotional communication, so that the apparent lower affective bandwidth appears to be an advantage for the display, expression and performance of more intense emotions. This shows the complexity of evaluating the affective bandwidth.

Forms of communication "colder" than face to face, with a lower affective bandwidth, are sometimes more suitable for emotionally charged messages, e.g. shy people, risk of losing face, embarrassment, etc. Face-to-face encounters are shaped by the face work and the self-control of the participants, technologically mediated encounters also have their forms of face work and self-control, but very often they are less constraining and allow for richer affective exchanges and for more affective information to be exchanged, as it was already acknowledged by the MUD users interviewed by Sherry Turkle more than a decade ago (Turkle 1995).

Regarding embarrassment, mobile media could be contributing to change our affective culture. The notion of affective culture provides a schema of experience and action, used by individuals to build their behaviour and to make sense according to their personal history, their style and their evaluation of the situation. The biological background is socially and culturally translated in ways that can be similar but also very different from one social group to another (Le Breton 1998) and also from one time to another, as Elias describes it. For instance, people in the Middle Age not only had different table manners and hygiene standards than today, they also had different relationships. These differences also affected their consciousness and affective economy. They did not know the invisible walls of affective reactions that separate our bodies. They did not experience the shame and embarrassment associated with the body and its physical functions that we know (Elias 1982). Mobile media practices reveal contemporary changes in affective culture about embarrassment associated with the body and self-disclosure of private and personal issues to micro-audiences of people we don't know (Lasén / Gómez 2009). This distribution of intimate images and accounts found in blogs, social network sites, social sharing sites, webs for personal contacts, or in IM exchanges between strangers, marks a current change about what is considered to be private and what is considered to be embarrassing. These images and narratives are the sign of "authentic emotions" (McQuire 2009: 196), which become a new social currency.

The assessment of the affective bandwidth has to consider not only the affordances and constrains of the particular media but also the asymmetry of the sender and the recipient. The choice of a so called colder channel in order to minimise the affective implication of the sender, as part of the face-work when communicating some bad news or when the message risks to harm or upset the receiver, can have the effect of making the receiving experience a much more emotional one. As the recipient's anger, frustration or sadness will be increased by the lack of consideration of the sender in the choice of the media to convey the message. For instance when a partner chooses to communicate the end of a love relationship by SMS or when an employer uses this same channel to let his workers know that they are made redundant[6].

The affective bandwidth is considered to play a role in the probability of the occurrence of misunderstandings. Therefore media with lower bandwidth could be fraught with a higher potential to elicit negative affective experiences. In our research, quoted above, about couples and mobile phones this is found in the statement of many participants who avoid arguing on the phone, by fear of misunderstandings, which could amplify the negative feelings. However this is not a problem for the long-term couples, which do not share this fear because they know each other well enough to be able to mutually interpret words and tone. This example reveals how the affective information conveyed by a media channel depends not only on the technical features of the media but as well on the persons who are using it and their experiences and the knowledge they already have about each other.

The particularities of writing media, such as email, SMS or IM, of being reviewed, replicable and forwarded, make them particularly suited to be shared and then submitted to collective interpretation (Myers 2007). Writing gives fluidity to discourses and turns knowledge into a flow, according to the Plato's view described in *Phaedrus*. Words lose experiential weight compared to orality and in their exteriority they acquire the ability to be shared, to be socialised, which become the support of collective memory, which afford the shift from individual memory, individual experiences, to collective memory. The fluidity and ability to be moved and mobilised by writing supports have not ceased to increase and accelerate through time till the widespread use of our mobile media screens (Stiegler 2009). The inscriptive particularities of mobile media support individual and collective reflexivity. They increase the affective potential of such media and the opportunities to elicit affective experiences as well.

Thus, the assessment of the affective bandwidth of a particular device or application has to take into account the particularities of affective information, the particularity of emotions. It is necessary to take into account the kind of affective experiences that emerged, in some cases unwillingly created by the sender intention of saving his or her face. And in any case, exceeding the intention and awareness of people involved in the exchanges, which raises questions about the ethical consequences of implementing devices and applications able to disclose, reveal and perform emotions. Hence, affective information could be considered as a kind of 'active information', borrowing Bohm's term (Clough et al. 2007), understood as a form of physical activity, which shapes, in-forms energy and matter.

6 In May 2003 British insurance company Accident Group fired 2400 workers, many of them received the announcement by an SMS sent by the human resources department to their company mobile phones, not during working hours, avoiding then a first collective reaction in the office. http://news.bbc.co.uk/2/hi/business/2949578.stm.

Conclusions

Emotions are an example of the shared agency between individuals and other entities. Being acted upon, put into motion, affected, can be considered as ways of sharing our personal agency. Nowadays technology uses and practices entail forms of shared agency between people and devices, where both entities are affected and mobilised by the other, where uses are the result of negotiations and sometimes clashes between technical affordances, commercial conditions and people's intentions, aims, habits and obligations, where non-intentional and non-conscious aspects are involved as well. This shared agency is a dynamic and learning process where all these aspects (affordances, conditions, intentions, habits, experiences, emergences) are subject to change and mobilised too.

The inscriptive power of mobile media, their ubiquitous presence, their ability to create affective occasions and to mediate emotional exchanges turn their practices and uses into privileged stages to observe and analyse the contemporary ambivalence and ambiguity of the "intensification of the self as the key site of human experience and its increasing sense of internal fragmentation and chaos" (Mansfield 2000: 2). These practices and uses become a particularly suited realm to put into practice a non-dualist approach to the study of emotions, accounting for the articulations, resonances and dissonances of their different aspects. Therefore, the assessment of the affective bandwidth of applications and devices has to take into account that affects and emotions present a particular kind of information, which is not reduced to meaning and bytes. Intensities, passions, encounters and experiences performed and elicited through mobile media, present complex power dynamics involving individuals, collectives, institutions and commercial interests. Therefore, the complexity of the affective implications of mobile media uses and practices demands a revision of the categories used to assess and develop their affective role.

References

Attwood, Feona (2009): 'deepthroatfucker' and 'Discerning Adonis'. Men and cybersex. In: International Journal of Cultural Studies, vol. 12(3), 279–294.

Boehner, Kirsten / DePaula, Rogério / Dourish, Paul / Sengers, Phoebe (2007): How emotion is made and measured. In: International Journal of Human-Computer Studies, 65 (2007), 275–291. URL: www.elsevier.com/locate/ijhcs (27-09-2009).

Boyd, Danah (2008): Facebook's Privacy Trainwreck. Exposure, Invasion, and Social Convergence. In: Convergence: The International Journal of Research into New Media Technologies, 14 (1): 13-20.

Clough, Patricia Ticineto / Halley, Jean O'Malley (2007): The Affective Turn: Theorizing the Social. Durham: Duke UP.

Clough, Patricia Ticineto et al. (2007): Notes towards a Theory of Affect-Itself. In: ephemera, vol. 7(1): 60-77. URL: www.ephemeraweb.org/journal/7-1/7-1cloughetal.pdf (30-09-2009).

Crawford, Kate (2008): These foolish things. On intimacy and insignificance in mobile media. In: Goggin, Gerard / Hjorth, Larissa (eds.): Mobile Technologies: From Telecommunications to Media. London: Routledge.

Critchley, Simon (2005): Things merely are. London: Routledge.

Damasio, Antonio (1994): Descartes' Error. Emotion, Reason and the Human Brain. New York: Grosset / Putnam.

Darwin, Charles (1872): The Expression of the Emotions in Man and Animals. New York: Philosophical Library.

De Graeve, Laurent (2001): Grégoire ou le téléphone portable. Brussels: Le Grand Miroir.

Deleuze, Gilles / Guattari, Felix (1987): A Thousand Plateaus. Capitalism and Schizophrenia. Translated and Foreword by Brian Massumi. Minneapolis: University of Minnesota Press.

Despret, Vinciane (2004): Emotional Makeup. Ethnopsychology and Selfhood. New York: Other Press.

Durkheim, Émile (1915\1965): The Elementary Forms of the Religious Life. New York: Free Press.

Ekmann, Paul / Friesen, Wallace / Ellsworth, Phoebe (1992): Emotion in the Human Face. New York: Pergamon.

Elias, Norbert (1982): The Civilizing Process. Oxford: Blackwell.

Featherstone, Mike / Hepworth, Mike / Turner, Bryan S. (eds.) (1991): The Body. London: Sage Publications.

Fortunati, Leopoldina (2002): The mobile phone: Towards new categories and social relations. In: Information, Communication & Society, vol. 5(4), 513–528.

Foucault, Michel (1988): Technologies of the Self. In: Martin, Luther H. et al. (eds.): Technologies of the Self: A Seminar with Michel Foucault. London: Tavistock, 16-49.

Franks, David D. / McCarthy, E. Doyle (1989): The Sociology of Emotions: Original Essays and Research Papers. London: JAI Press.

Gardner, Howard (1973): Frames of the Mind. NY: Basic Books.

Gilbert, Jeremy (2004): Signifying Nothing: 'Culture,' 'Discourse' and the Sociality of Affect. In: Culture Machine, vol. 6. URL: /www.culture machine.net/index.php/cm/article/viewArticle/8/7 (28-09-2009).

Goggin, Gerard / Hjorth, Larissa (eds.): Mobile Technologies: From Telecommunications to Media. London: Routledge.

Goleman, Daniel (1997): Emotional Intelligence. New York: Bantam.

Green, Nicola (2001): Who's watching whom? Monitoring and accountability in mobile relations. In: Brown, Barry / Green, Nicola / Harper, Richard (eds.): Wireless World. Social and Interactional Aspects of the Mobile Age. London: Springer-Verlag, 32-45.

Grossberg, Lawrence (1992): We Gotta Get Out of This Place – Popular Conservatism and Postmodern Culture. New York: Routledge.

Grosz, Elizabeth (1994): Volatile Bodies. Towards a corporeal feminism. Bloomington: Indiana University Press.

Haraway, Donna (1991): A Cyborg Manifesto: Science, Technology, and Socialist-Feminism in the Late Twentieth Century. In: Simians, Cyborgs and Women: The Reinvention of Nature. New York: Routledge, 149-181.

Hardt, Michael (n. d.): "Affective Labour". URL: www.vinculo-a.net/english_site/text_hardt.html (07-09-09).

Hjorth, Larissa (2008): Mobile Media in the Asia-Pacific. London / New York: Routledge.

Hochschild, Arlie R. (1979): Emotion work, feeling rules and Social Structure. In: The American Journal of Sociology, (85), 551-75.

Hochschild, Arlie R. (1983): The managed heart: Commercialization of human feeling. Berkeley, CA: University of California Press.

Kemper, Theodore D. (1991): An Introduction to the Sociology of Emotions. In: International Review of Studies on Emotion, vol. 1, 301-49.

Kücklich, Julian (2005): Precarious Playbour: Modders and the Digital Games Industry. In: Fibreculture (5). URL: http://journal.fibreculture.org/issue5/kucklich.html (28-08-2009).

Lasén, Amparo (2001): Le temps des jeunes: rythmes, durée et virtualités, L'Harmattan, Paris.

Lasén, Amparo (2004): Affective technologies: emotions and mobile phones. In: Receiver, vol. 11. URL: http://www.receiver.vodafone.com/11/articles/index03.html.

Lasén, Amparo (2005): Understanding Mobile Phone Users and Usage. Newbury: Vodafone Group R&D.

Lasén, Amparo (2009): 'Mobiles are not that personal': the unexpected consequences of the accountability, accessibility and transparency afforded by mobile telephony. In: Ling, Rich / Campbell, Scott (eds.): The Mobile Communication Research Series: Volume II, Mobile Communication: Bringing Us Together or Tearing Us Apart? Edison, NJ: Transaction books.

Lasén, Amparo / Gómez, Edgar (2009): Digital Photography and Picture Sharing: Redefining the Public/Private Divide. In: Knowledge, Technology and Policy, vol. 22 (3), 205-215.

Latour, Bruno (2004): How to talk about the body. The normative dimensions of science studies. In: Body and Society, 10 (2-3): 205-229.

Lazzarato, Maurizio (2004): 'From Capital-labour to Capital-life'. In: Ephemera 4: 187-208.

Le Breton, David (1998): Sociologie des emotions: Critique de la raison darwinienne. In: Recherches sociologiques 1, 37-54.

Lutz, Catherine / White, Geoffrey M. (1986): The anthropology of emotions, Annual Review of Anthropology, vol. 15, 405-436.

Maffesoli, Michel (1993): The Shadow of Dionysus: A Contribution to the Sociology of the Orgy State. New York: University of New York Press.

Maffesoli, Michel (1996): The Time of the tribes: The Decline of Individualism in Mass Society. London: Sage.

Maidment, Jane / Crisp, Beth (2007): Not just for romance: Applications of 'speed dating' in social work education. In: Groupwork An Interdisciplinary Journal for Working with Groups, Volume 17(2), 13-27.

Mansfield, Nick (2000): Subjectivity. Theories of the Self from Freud to Haraway. New York: New York University Press.

Massumi, Brian (2002): Parables for the Virtual – Movement, Affect, Sensation. Durham: Duke University Press.

McCarthy, E. Doyle (1989): Emotions are Social Things: An Essay in the Sociology of Emotions. In: Franks, David D. / McCarthy, E. Doyle (1989): The Sociology of Emotions: Original Essays and Research Papers. London: JAI Press, 51-72.

McQuire, Scott (2008): The Media City. Media, Architecture and Urban Space. London: Sage.

Mørk Petersen, Søren (2009): Common Banality: The Affective Character of Photo Sharing, Everyday Life and Produsage Cultures. PhD Thesis, Department of Innovative Communication at the IT-University of Copenhagen, Denmark.

Myers, Piers (2007): Sexed up intelligence or irresponsible reporting? The interplay of virtual communication and emotion in dispute sensemaking. In: Human Relations, 60, 4; 609-636.

Norman, Donald A. (2003): Emotional Design. Why We Love (or Hate) Everyday Things. New York: Basic Books.

Pertierra, Raul (2005): Mobile phones, identity and discursive intimacy. In: Human Technology, Volume 1 (1), 23-44.

Picard, Rosalind W. (1997): Affective Computing. Cambridge: The MIT Press.

Picard, Rosalind W. (2003): Affective Computing: Challenges. In: International Journal of Human-Computer Studies, Volume 59, Issues 1-2, 55-64.

Picard, Rosalind W. / Papert, Seymour / Bender, Walter / Blumberg, Bruce / Breazeal, Cynthia / Cavallo, David / Machover, Tod / Resnick, Mitchel / Roy, Deb / Strohecker, Carol (2004): Affective Learning - A Manifesto. In: BT Technical Journal, Volume 22, No. 4, 253-269.

Prøitz, Lin (2005): Intimacy Fiction Intimate Discourses in Mobile Telephone Communication amongst Norwegian Youth. In: Nyiri, Krystóf (ed.): A sense of place: The global and the local in mobile communication. Wien: Passagen Verlag, 191-200.

Qiu, Jack Linchuan (2007): The Wireless Leash: Mobile Messaging Service as a Means of Control. In: International Journal of Communication 1, (2007), 74-91.

Reeves, Byron / Nass, Clifford (1996): The Media Equation. How People Treat Computers, Television and New Media like Real People and Places. New York: Cambridge University Press/CSLI Publications.

Reynolds, Carson J. / Picard Rosalind W. (2004): Affective Sensors, Privacy, and Ethical Contracts. In: Proceedings of 2004 Conference on Human Factors and Computing Systems (CHI 2004). Vienna: ACM Press.

Reynolds, Carson / Picard, Rosalind W. (2005): Evaluation of Affective Computing Systems from a Dimensional Metaethical Position. In: 1st Augmented Cognition Conference, in conjunction with the 11th International Conference on Human-Computer Interaction, July 22-27, 2005, Las Vegas, NV.

Rosaldo, Michelle Z. (1984): Toward an anthropology of self and feeling. In: Shweder, Richard A. / LeVine, Robert A. (eds.): Culture Theory: essays on mind, self and emotion. Cambridge, UK: Cambridge University Press, 137-157.

Ruin, Hans (2000): The Passivity of Reason: On Heidegger's Concept of Stimmung. In: Sats - Nordic Journal of Philosophy, vol 1 n 2, 143-159.

Scherer, Klaus R. (2001): Emotional experience is subject to social and technological change: extrapolating to the future. In: Social Science Information, 40 (1), 125-151.

Sennett, Richard (1977): The Fall of the Public Man. New York: Alfred A. Knopf.

Shouse, Eric (2005): Feeling, Emotion, Affect. In: M/C Journal 8.6 (2005). URL: http://journal.media-culture.org.au/0512/03-shouse.php (10-09-2009).

Siegel, Daniel J. (1999): The Developing Mind: Toward a Neurobiology of Interpersonal Experience, New York: Guilford Publications.

Sloterdijk, Peter (1989): Thinker on stage: Nietzsche's materialism. Minneapolis: University of Minnesota Press.

Stiegler, Bernard (2009): Technics and Time, 2: Disorientation. Stanford: Stanford University Press.

Thrift, Nigel J. (2006): Re-inventing Invention: New Tendencies in Capitalist Commodification. In: Economy and Society 35: 279–306.

Tomita, Hidenori (2006): Keitai and the Intimate Stranger. In: Ito, Mizuko / Okabe, Daisuke / Matsuda, Misa (eds.): Personal, Portable, Pedestrian Mobile Phones in Japanese Life. Cambridge: MIT Press, 183-20.
Turner, Bryan (1984): The Body and Society. Explorations in Social Theory. Oxford, UK: Blackwell [2nd Edition published by Sage, 1996, London].
Turkle, Sherry (1995): Life on the Screen: Identity in the Age of the Internet. New York: Simon & Schuster.
Vincent, Jane (2005): Emotional Attachment and Mobile Phones: An Extraordinary Relationship. In Hamill, Lynne / Lasén, Amparo (eds.): Mobile World: Past, Present and Future. London: Springer, 95-104.
Vincent, Jane (2006): Emotional attachment and mobile phones. In: Knowledge, Technology, and Policy, Volume 19, Number 1, 39-44.

Living with mobile phones

Jane Vincent

Introduction

Mobile phones are now omnipresent in the UK with some 86 % of UK adults and slightly fewer children aged 12-15 owning at least one (Ofcom 2008). Each year a diminishing number of people never use them with most adults and children of all ages having access to a mobile phone even if they do not own one. In addition to the increased use of mobile phones for voice calls, during 2008, an average of 218 million text messages were sent each day in the UK of which 1.5 millions were picture and video messages (Mobile Data Association 2009). People use mobile phones in many different ways, assisted by favourable prices for accessing the internet and other convergent services. This is exemplified by this quote from the UK Office of Communications (Ofcom 2008) report published in August 2008:

> "Mobile handsets are increasingly convergenced devices, with 41 % of those with a mobile phone using them to take photos, 17 % using them for games and 13 % making use of their diary / organiser functionality" (Ofcom 2008: 113).

This chapter explores the ways that some of these mobile phone users in the UK have appropriated mobile phones and integrated them into their lives. In particular it examines how mobile phones may have affected their social and familial practices such as those involving assurances of safety, making arrangements, relationships and work practices. It also discusses the role of the mobile phone in the electronic emotion and indeed, the development of the electronic self of each and every mobile phone user. It draws on a series of empirical research studies conducted for the Digital World Research Centre[1] (DWRC) since 2002 and focuses on the author's most recent research that examined the use of mobile phones by 40 respondents in the UK aged over 40. The various DWRC studies discussed in this chapter are linked by similar qualitative research methodologies with the data being obtained in a variety of ways primarily from one to one interviews but also 24 hour diaries, focus groups and questionnaires. The data is explored with particular reference to the work of the interactionist theorists and is informed by mobile communications studies of

---

1 Digital World Research Centre Faculty of Arts and Human Sciences University of Surrey UK www.dwrc.surrey.ac.uk.

others during the same research period such as that of Höflich (2003) and Fortunati et al. (2003).

Theoretical discourse

Using mobile phones requires social, physical and sensorial interaction between humans and these small electronic computational devices (cf. Vincent 2006; Katz 2003; Fortunati et al. 2003). Although mobile phones have been in use since the late 1970's, academic studies that explored the ways people incorporated them into their day to day lives were scant until the early twenty first century by which time mobile phones had become more prevalent worldwide. Exceptions were studies by the COST group (Haddon 1998; Fortunati 1998; Ling 1998). Theoretical analysis and the development of new social theory to explain this burgeoning phenomenon were also scarce as researchers struggled to obtain data and examine the ways that mobile phones were being used. Research conducted by organisations within the mobile communications industry remains largely inaccessible to academia or, similar to technological and engineering research, it deferred to development of international telecommunications standards for guidance on user requirements and experience as in the work of the International Telecommunications Union (ITU) and later the European Telecommunications Standards Institute (ETSI) (Tafazzoli 2006; Goggin 2006). Additionally a few studies have been published by industry organisations such as the UMTS Forum and the GSM Association, including those carried out by the DWRC for the former (Vincent / Harper 2003; Vincent / Haddon 2004).

In contrast with this worldwide paucity of research on the adoption of mobile phones the effect of the introduction of new media on established communications social practices has been the subject of research for many decades. For example, Short et al. discussed their concept of the 'social presence' of a communications medium with regard to fixed telephony and how this was affected by: […] 'the capacity to transmit information about facial expression, direction of looking, posture, dress and non-verbal vocal clues' (Short et al. 1976: 65). They also refer to the work of Argyle and Dean (1965) on intimacy that explored the concept of the equilibrium that two people seek when they enter into a conversation. This equilibrium, they asserted, was upset by the intervention of a "third party" communications medium. Communications, linguistics, psychology, neurophysiology and sociology are among the many disciplines which now explore the use of mobile phones but the development of a theory that goes some way to explaining the extraordinary phenomenon of mobile phone use still appears to elude the social sciences.

In addition the seminal works of McLuhan (1964) and Meyrowitz (1985), written without reference to mobile phones, were influential in early studies about the mobile phone as researchers compared and contrasted the adoption, and communication behaviours with these other, less mobile, information communication technologies (ICTs). Whilst there are indeed similarities with the mass adoption of these ICTs a particular difference is how the mobile phone is used as a one to one communications device and rarely shared within a household. In contrast the radio, television and even the fixed line telephone were shared within households, with communications on radio and television being targeted at a mass audience rather than a small group known personally to each other as is the situation with mobile phones. However, the use of a mobile phone in public, only made possible after the introduction of 'hands free' pocket or handy phones in the 1990's, brought to the fore the issue of clashes between public and private behaviours already encountered with other electronic mediums, and illustrated by this quote from Meyrowitz:

> "Many Americans may no longer seem to "know their place" because the traditionally interlocking components of "place" have been split apart by electronic media. Wherever one is now – at home, at work or in a car – one may be in touch and tuned in" (Meyrowitz 1985: 308).

These observations and examinations of place and its affect on peoples' sense of their self were already topics of theoretical discourse by Goffman (1959) in particular who is widely cited by leading researchers in the field of mobile communications and studies of electronic communications. Ling notes that Goffman explored the use of the telephone, albeit briefly, and that in his early writings he saw the telephone as being used only privately. However, in later writings, he considered the use of telephones in more public spaces and commented on the behaviour of those co-present. Ling identifies some clear analogies with mobile phone behaviours in the observations of Goffman on the use of fixed line telephones such as the way that the co-present change their behaviour when a telephone call is answered allowing the recipient of the call (and thus the telephone as well) to be the dominant party. However, as Ling notes, the role and use of the telephone prior to the advent of mobile phones was not the same as in the present day when seemingly everyone has their own personal phone and has no need to share a public phone booth, household phone, or dormitory phone (Ling 2008: 69).

There have been some attempts to address this problem of a social theory of mobile phone use. The domestication theory of Silverstone (Silverstone / Haddon 1996) has been used by some (Haddon 2003; Hartmann et al. 2005) in so far as the adoption of the device and the role of the mobile phone in the social aspects of everyday life was found to have similarities with the explanations for the adoption of other ICTs, but this only accounts for a small part of the ways the mobile phone has become so embedded into people's lives beyond the home.

During this first decade of the twenty first century mobile communications research has burgeoned and a great deal of the theoretical discourse associated with it refers to the interactionists, and, as indicated above, to Goffman in particular. Some have also brought the concept of absence presence to bear on the debate (Gergen 2002; Fortunati 2002) reflecting the ways the mobile phone can appear to bring people together even when apart as if they are in the same place, a place created when using the mobile phone. Regardless of the distance or reason for separation the mobile phone has a role in overcoming the absence of others developing what Gergen refers to as:

> "[...] the growing domain of diverted or divided consciousness invited by communications technology, and most particularly the mobile telephone. One is physically present but absorbed by a technologically mediated world of elsewhere" (Gergen 2002: 227).

This concept of being subsumed by the technology of the mobile phone is powerfully represented in the lives of those who have migrated from their home but maintain strong familial ties and responsibilities from afar. As Pertierra writes in his research on Philippine diasporas:

> "Mobile phones allow absent subjects to exercise a daily presence in their communities of origin. This absent presence generates virtual subjects interacting primarily via the mobile phone" (Pertierra 2005b: 23).

The effect of being with the absent present also affects the co-present during the communication. Höflich researching the disruption caused by mobile phone use in a public place (Höflich 2009) observes how people continue to walk and talk in this public place but are seemingly in another place altogether whilst doing so. When they complete their call they return to being with the co-present although they have physically been in the same location the whole time.

This ability to conduct mobile phone communications almost anywhere raises some interesting points with regard to human interaction. As stated earlier, mobile phones appear to have social, physical and sensorial effects. It is considered by sociologists writing today that reference to face to face in the context of communication has been supplanted by body to body experiences (Fortunati 2005; Pertierra 2005a), and indeed this is borne out by the experience of using a mobile phone. The mobile phone requires the user to employ aural, oral, and tactile senses as well as invoking smell and taste in the memory and conversational experiences they engender (Vincent 2006). Building on the work of Maldonado who discusses the body and how technology is developing ways to emulate it, it is as if the mobile phone is piercing the body with this assault on the senses demanding multiple reactions that over-ride other simultaneous actions. As Maldonado says:

> "One very important point is usually overlooked. A person's natural sense of touch does not consist only of contact, touching is not just touching. Our sense of touch perceives multiple factors even without true direct contact with our skin" (Maldonado 2003: 20).

It is perhaps no surprise therefore that the use of a hand held mobile phone whilst driving a vehicle is considered such a hazard (and punishable as a criminal offence) in the UK. Compared to the use of other computer mediated electronic devices such as satellite navigation, radio or music player, the mobile phone has the added involvement of body to body human interaction.

This body to body effect of the mobile phone has been particularly notable with regard to the emotion it invokes and the relationships it strengthens and this brings one again to the work of the interactionist theorists, and in particular the theory of emotion labour developed by Hochschild (1983, 2003). Although this theory is mostly used to examine the roles of workers in professions such as health or customer care the principle of emotion labour can also apply to human interaction via a mobile phone, not only by these professions but also by everyday users. Hochschild, in a recent paper highlights this emotion link when dealing with people via the telephone across continents; they may never meet but are nevertheless closely involved in personal and private matters of social life:

> "[..] Emotional labor crosses borders in other ways as well. Through telephone and email, service providers in Bangalore, India, for example, tutor American children with math homework, make long-distance purchases of personal gifts, and even scan romantic dating service Internet sites for busy professionals. What we see here are the paradoxes – and sometimes estrangements – involved in commodifying even the smallest, most personal acts" (Hochschild 2008).

Hochschild has also highlighted what she has termed 'moments of pinch'; the point at which one's own emotions are under most pressure from a situation in which they cannot be expressed. This conflict of wanting and not being able to express oneself creates a heightened emotion which can make handling the situation even more tense. There are some examples of this discussed later in this chapter.

Having considered theories applicable to examining mobile phone use the matter of developing theory and in particular that which explores the concept of electronic emotion and the electronic self is now explored. The role of electronically mediated communications in our everyday lives (and certainly within the lives of the respondents discussed in this chapter) has become so imbued in our body to body relationships that it may elicit new 'electronic emotion'. This emotion is made possible and perhaps felt only when the mobile phone participates as mediator, be it as an actual conduit for the electronic emotion or merely the prompt or cause of it by simply holding or thinking of a particular mobile phone. This concept of electronic emotion of e-actors (cf. Vincent / Fortunati 2009c) communicating body to body experiences is still in its infancy and one that is being examined further. However, what this new research does emphasise is how the mobile phone appears to be much more than a device that enables constant and immediate connectivity. Indeed this desire not to be switched off, not to lose the connection has created the notion of a 'dependent presence' between those who are co-present as well those who are apart – the absent present. This is a corollary to the states of absent and co-presence that Gergen, Fortunati and others, discussed earlier, have explored. The dependent presence occurs as a result of needing and wanting to feel together, to share emotion (electronically), and the ability to be always touching or feeling the mobile phone sates this for some.

The next section of this chapter explores, through examination of interviews from the author's most recent research, examples of emotion labour, electronic emotion and indeed this dependent presence, the increasingly sensorial body to body experience that is brought about by the mobile phone and that is manifest in the electronic self. In these examples of the use of a mobile phone discussed below there is an added dimension of intimacy and presence in the way it enables this sense of being together, of being in the same place when you are not. These respondents' lives, their memories, relationships and their emotion, have become intimately interwoven into the device itself, an outcome of the continuous, prolonged and strengthening intimacy achieved through the mobile phone's mediation of electronic emotion.

Exploring the data

The various studies into the social practices of mobile phone users conducted over a period of 7 years by the author and her colleagues (Vincent 2009a, 2009b) at the Digital World Research Centre have identified some consistent and enduring themes. These have been used to inform the research questions developed for each subsequent study. In this chapter two of these themes form the basis for the exploration of research topic discussed, namely the social, physical and sensorial aspects that appear to be imbued in the mobile phone. In this section I will expand upon some of the respondents' accounts of their mobile phone use to explore this research topic in more detail. The first of these two themes is the ways that mobile phones are used for strengthening already existing social contacts and thus used to build relationships – private and business – as well as to make arrangements, many at short notice. Secondly, it was found quite early in the research that people appear to have an emotional relationship with their mobile phone that is different from that with any other electronic communications device (Vincent / Harper 2003; Vincent / Haddon 2003). It is these two themes that I will use as the basis for my examination of the research, most particularly the most recent research conducted during 2007-

2008 on people aged over forty. These latter respondents, unlike the children aged 11-16 in my earlier studies (Vincent 2004; Haddon / Vincent 2009), have each experienced the use of telecommunications services and other information and communications technologies in their adult life before the public availability of mobile phone services. Notwithstanding this, some of the respondents have actually used a mobile phone for over 20 years having had them in their employment in the mid to late 1980's. One respondent recalled what it was like to be one of the few with a mobile phone in those days:

> "I suppose I am quite unusual in that I have used mobile phones right from the start. The first call I made on a hand portable was on Holborn Viaduct railway station, [...] which was in Easter 1985, and I made a call to tell my wife I was coming home, at the time she was a housewife at home, and I was looked upon by people at the station carrying this £3000 Motorola 8000x brick as if I was somebody who lived on another planet" (Simon).

Most, however, have only had a mobile phone for less than 10 years acquiring them when they became more easily and cheaply available. *"I just decided it was really a good thing to have, we did a lot of weekends away [...]"* (Bart). For others it was after a particular incident when having one would have been useful such as Elizabeth and her husband who bought their first mobile phone after they had broken down on the way to hospital when Elizabeth was having a baby and had no easy way of calling for help. Leanne, on the other hand bought one *"because everyone else had one"*. Some of the respondents were reluctant adopters of mobile phones and one, Sara, had tried to find a reason why mobile phones seemed to dominate and figure so highly in her life and that of her contemporaries by comparing this with the way some younger friends behave with their phones.

> "Mobiles do tend to have this priority over everything, I don't know whether it's our age group; [...] I do wonder whether it's because it is new technology to us and we were older, whether we let it take over our lives; they [the 18 year olds] have grown up with it and it's a part of their lives. There is a big distinction there" (Sara).

Unlike Sara's wariness of being dominated by the mobile phone, some respondents liked the gadgetry and technology, and this was also partly the reason why June had bought a mobile phone.

> "I've just got the one phone; I got a new one in the summer time; I'm actually very, very attached to my phone. It's a very important part of my life. And I love playing with it; it's a wonderful gadget that I love playing with" (June).

In the UK, technological artefacts are known as 'gadgets' and many of the respondents referred to their mobile phone as a gadget, an example, perhaps of the domestication of the product within the portfolio of the various technologies they use on a daily basis. Another respondent, Gordon was given his first phone as a present from his wife who knowing he liked gadgets thought he would like a mobile phone. She did not get one herself for another two years but they both use them to keep in contact with each other now. Jeremy – who had his first mobile in 1987 – was thrilled to have just received a new one on loan from a business contact, it was the latest Blackberry device and he was enjoying seeing what it could do, contrasting it with his last model that had broken. Regardless of whether they like or dislike the phone as a technology artefact, owning and using a mobile phone for these respondents had become a routine, even mundane, requirement of their daily life and most of them were never unintentionally without it, although a few found they were expected to always have it with them. Helen's new fiancé, Jim, used his mobile phone constantly and she had discovered he expected the same from her:

> "I do remember a time when I left my phone at home and Jim was trying to contact me, and it transpired he couldn't, and that's happened a couple of times, and because his phone is with him constantly he couldn't understand why he couldn't get hold of me and when I did actually get through to him I just said, look I just didn't have my phone, I just didn't have it, but it was an issue" (Helen).

Asked if this was because he was worried about her or because he was cross she did not have it with her she said: *"I think it was more cross, more cross than worried, or couldn't understand, he couldn't understand why I wasn't communicating with him"* (Helen). This expectation that people would always have their phones with them was reported in numerous interviews and as June said: *"I can't think of anyone who hasn't got a mobile phone. Most of them are very comfortable with that modus operandi and therefore have their mobile phones".* Thus, whether or not these respondents used their phones a great deal, even carrying it around in their hand to avoid missing a call, or they only switched it on when they needed to make a call, the respondents found they had come to need their mobile phone, not least because friends and family and business contacts expected them to use it too. For some this was ruefully commented upon, especially if it was international clients or their boss who expected them to be available twenty-four hours a day seven days a week.

> "I think it's completely changed the way we live really, this whole 24by7 being able to contact anybody anywhere anytime. It's changed the way we do things; you don't have to be organised in the same way, you can do things differently, you don't have to rely on phone boxes working to get home and you just get used to being able to contact people anywhere anytime and that's a big change – I'm not necessarily convinced it's a good change, it's just part of the overall speeding up of the pace of life" (Mike).

Most of the respondents had experienced moments, some for several days, when they had been unexpectedly separated from their mobile phone due it being broken or mislaid. The examples they gave of what bothered them most when this happened was in each case about loss of contact with loved ones and, for some, the unexpected feeling of being isolated. This despite the fact that they had until that moment been talking mostly about how important the mobile phone was to help them deal with organisational aspects of their life such as emails and business correspondence. Carl spoke about the temporary loss of his Blackberry on an overseas business trip, he had quickly realised where he had mislaid his mobile phone and went to great effort to recover it. However, although it was later returned to him he was left without it for a few days. He spoke of this time not in terms of the difficulties of keeping in contact with work colleagues, a problem that could be overcome with a few phone calls but instead it was the loss of immediate and instant contact with his family whilst he was so far from home that upset him most. It made him feel very alone particularly when he was in his hotel room late at night and could not just quickly text good night, or have a chat to his wife as he would normally have done, or look back at a few texts and pictures of his friends and family.

Some respondents appeared a little discomfited to admit their dependence on the mobile phone having played down its importance to them, reflecting as the interview progressed that: […] "*of course I do use it ...*"(Nina), when they had thought they would just have it for those emergency moments. Whilst her husband was very ill and housebound Nina had found she was able to pop out for an hour for essential shopping knowing he could call her to come home if she was needed. Nina spoke of the importance of the mobile phone and how she would even take it into the garden with her or when she was in a different part of the house so she could always be reached in an emergency. It was never used but knowing that she had it as she moved out of audible contact reassured both of them.

As they talked it was as if some respondents did not want to admit how much they did rely on their mobile phone or how much others, usually their partner / spouse or child, expected them to always have it with them. This was not only for safety and security reasons but for fun too. For example on a football trip abroad when England had performed particularly well the phone was *"red hot all the way back to the house in the car"* as Charles shared the experience of the match with his son who had watched it on the television but could not be with him in the stadium.

Whether or not the mobile phone had been purchased for safety and security reasons there was little doubt that all the respondents used it in this way and furthermore this mode of use extended to those for whom they were responsible whether they still lived at home or not. Keith recounted how he rued the day he forgot to take his mobile phone with him when, having left his daughter at the station to catch a train back to university, he would have liked to text or phone her just to make sure the train had turned up. Stephen talked about how his young daughter uses her mobile phone to call her mother at home as she walks home from the bus stop on the way home from school; this reassures both his daughter and his wife. June also talked about the importance of having a mobile phone with regard to the unpredictability of her pre-school children:

> "[...] Because with children, they are quite unreliable things and you tend to be late leaving or late arriving or whatever. It's very handy to have the convenience of being able to just phone up and say we're late, we're here, we're lost or whatever; really, really helpful to have that, it's important really to have a mobile phone" (June).

When asked in the interview *"how many mobile phones do you have?"* several respondents said, *"what me or in our house?"*. Although a personal device, mobile phones had in some instances initially been shared within households (a reflection perhaps of the way mobile phones were used as a shared resource in offices in the 1980's, as recalled by Carl, and maybe the fixed phone too). The role of the mobile phone in households was much more than simply an organising tool although it seemed this was its guise rather than an explicit role, as there did not appear to have been any agreement to use the mobile phone to 'keep tabs on' each other although this was how it was being used. Mike recalled how the mobile phones are used when his family go on foreign holidays, he, his wife and 2 teenage children each have a mobile phone and they use it then as a means of pursuing their individual interests.

> "I think it's a very useful tool. We often split up. We want to look round bits of touristy stuff and the kids want to look round the shops or Andrew wants to go and stand in a computer shop for an hour playing games rather than see the exotic touristy stuff. We make loose arrangements to meet up in an hour or so and then confirm them by phone" (Mike).

However, keeping tabs on each other was more than a time management or diary exercise, it was a means of staying in touch, by maintaining a constant electronic presence, they felt together and that they were safe. As found in the earlier studies this constant use of the mobile phone also strengthened these relationships. Maria kept her mobile phone in her pocket when she could not hold it in her hand as she could feel it when she received a communication. She talked about her anxiety when she lost weight and her trousers became loose, reporting what she had said to her boyfriend when he complained she had not answered a call: *"Yes but my trousers are loose what do you want me to do? Hold it in my hand all the time?"* – Maria admitted that she did now hold the mobile phone in her hand. Maria's relationship with her boyfriend, (who lived in a different country), was quite volatile and the mobile phone played a central role not only as the mediator of all their communications but also as the repository of her joy, angst and anger in messages and images stored and sent, as well as being thrown down on her bed in frustration when her boyfriend refused to answer her calls. Their desire to remain in constant contact with each other through texts, multi-media messaging and phone calls is demonstrative of the notion of dependent presence discussed above. Maria and her boyfriend needed to feel the presence of the other at all times but this had led to Maria returning home early when she accompanied a friend on holiday. As she explained:

> "[...] she got annoyed – this is interesting – because of my ... because I was texting him quite a bit and at some point I asked to excuse myself and have a half hour conversation with him and she got annoyed about that and she said you're not here with me you're there with him and I want you to leave tomorrow" (Maria).

This sense of dependent presence between another respondent and her daughter is augmented by the use of the social networking site Facebook.com. Sally had realised her daughter's activities at university were appearing on her Facebook site more quickly than she would find out directly from her. So, with her daughter's knowledge, she checks on her Facebook then texts her daughter or her friends to find out more about what she has been doing. She had also downloaded and printed some photos of her daughter with her friends they had taken with their mobile phones for the Facebook site. She had a close relationship with her daughter who appeared happy to involve her mother in this way during the transition of leaving home. Sally maintained an almost constant co-presence with her daughter using the mobile phone and the Internet. She saw this as a perfectly normal mother / daughter relationship and indeed implied it was strengthening the bond that was put under strain after the daughter moved away from home for the first time to go to university. This manifestation of the dependence presence also occurs in night-time calls when her daughter phones her mother after a row with her boyfriend to seek solace in her mother's voice and her counsel.

Bart and Rita use their mobile phones to talk to and text each other daily and in their interviews (conducted separately) both talked about how they use it to meet up when shopping, to find out what is happening or just to stay in touch by text. To begin with, when the children were small, they shared one mobile phone and Rita would take it with her when she was with their children and staying away at her mother's. Now, they each have a mobile phone as well as their older son who is now 16, although their younger son, aged 13, has use of a family back up phone when it is essential for him to have a mobile phone. This latter point is recognition of a need but also a response to Bart and Rita expecting their sons to be accountable and responsible for how much they used their mobile phone; they had found to their (financial) cost that their younger son was not yet ready to take this responsibility. The use of mobile phones as a constant reminder of happy memories, a commonly reported use, is exemplified by Helen who uses a special picture as wallpaper on her mobile phone to remind of when her fiancé proposed:

> "I've got Chichester Cathedral on there at the moment because that's where we went when we got engaged – at Chichester so that's quite nice" (Helen).

Lucy recalled times when she wanted to talk to her husband so she had simply rung him without particularly thinking of his whereabouts realising only later that he was actually in a church service and could not be relied to turn off the sound or turn off his mobile phone. Nowadays she is more circumspect about when and where he might be when she calls him and so might use text instead but she does still make contact whenever she needs to.

This diversity of use and the reasons given for having the mobile phone emphasised the actual need or indeed dependence on the mobile phone that mirrors the dependence these respondents had on other family members. The mobile phone would also appear to satisfy the need to sense the presence of others through handling, feeling or even thinking about the device, the text or the image. Whether they used it a few times, hardly at all or constantly, the mobile phone was a necessity for all these respondents. In common with others one respondent was incredulous when asked if she could live without a mobile phone: "*What – if mobiles didn't exist? I can't use it at all?*" (Maria).

The use of the mobile phone to mediate a dependent presence was also manifest in maintaining memories and feelings with family and friends who have died. Although he did not do so in haste as he felt that would be disrespectful, Jeremy removed all details of a deceased relative for fear of accidentally making contact via a mistaken call or text because the data was still stored on his phone. In this instance his intentions were more about avoiding offence to others rather than dealing with his own feelings. However, in contrast deleting someone from the mailbox and deleting their messages was a hard decision for some and a significant part of the process of closure in the grieving of a loved one. Carl had lost a close friend and although he had managed to delete his email addresses and contact details on his computer, even two years on, he could not bring himself to delete the phone number or his friend's last text messages. Because he could hang on to them it was: "*[...] as if he hadn't completely gone*" (Carl).

Thus the notion of dependent presence can apply to relationships with people no longer alive as well as those who have moved away or who might be close at hand, such as assisting in the execution of parental responsibilities and in keeping husband and wife or partners in constant contact. The mobile phone in each instance serves as a conduit between people delivering an electronically mediated memory even when there is no prospect of ever physically communicating again. The touch of the mobile phone to re-read messages and the text as a visual record of their contact each acts to keep the sensorial and familial aspects of the relationship alive.

Conclusions

In this chapter the two themes – strengthening and managing relationships and the emotional relationship people have with their mobile phone – have been examined through the data obtained from recent research, and with reference to previous studies. These themes manifest in many ways as examined above in the accounts of the respondents' experiences and has been explored through the notion of dependent presence. This notion is apparent in the respondents' dependence on mobile phones for safety and to keep in touch with loved ones, their delight in the technology, and the emotion they express when they talk about and use their mobile phone. The reasons for using the mobile phone were also explored such as to set up social arrangements, to stay in contact at all times when separated, to remind one of happy times by handling the device and looking at images, and simply to make one feel that the other absent person is present even when they are thousands of miles apart, or even no longer alive.

Examining the ways these respondents used their mobile phones has highlighted the social, physical and sensorial aspects of the body to body interaction that they enable. Most respondents relied on the contact their mobile phone makes possible with close friends and family as well as allowing them to maintain the history of this kinship and familial contacts. The combination of the mobile phone and its communicants extends beyond this to include work and business contacts and in total it represents a repository of multiple mediated relationships in which the electronic self is stored alongside the electronic emotion associated with it.

Indeed, the mobile phone has become, for some respondents, an inviolable place containing intimate data and meanings that are a unique outcome of the electronic interplay between the owner of that device and their correspondents. Thus it is not only the actual material in the form of phone numbers, text messages, images and the personal set up of the device that is important here, but rather the whole meaning the mobile phone has for its owner as a result of the history of its use and the relationships that surround and complement its communicants.

References

Argyl, Michael / Dean, Janet (1965): Eye Contact, Distance and Affiliation. In: Sociometry 28, 289-304.

Fortunati, Leopoldina (1998): The Ambiguous Image of the Mobile Phone. In: Haddon, Leslie (ed.): Communications on the Move: The Experience of Mobile Telephony in the 1990s, COST248 Report, 121-148. Telia AB, Fasta. URL: http://www.cost269.org/.

Fortunati, Leopoldina (2002): The mobile phone: towards new categories and social relations. In: Information, Communication & Society, Vol. 5, 4, 513-528.

Fortunati, Leopoldina / Katz, James E. / Riccini, Raimonda (eds.) (2003): Mediating the Human Body: Technology, Communication and Fashion. Erlbaum: Mahwah, New Jersey.

Fortunati, Leopoldina (2005): Is Body-to-body communication still the prototype? In: The Information Society 21(1): 1-9.

Gergen, Kenneth J. (2002): The Challenge of Absent Presence. In: Katz, James E. / Aakhus, Mark (eds.): Perpetual Contact Mobile Communication, Private Talk, Public Performance. Cambridge: Cambridge University Press, 227-241.

Gergen, Kenneth J. (2003): Self and Community in the New Floating Worlds. In: Nyiri, Kristóf (ed.): Mobile Democracy: Essays on Society, Self and Politics. Wien: Passagen Verlag, 103-114.

Goggin, Gerard (2006): Mobile Phone Culture. London: Routledge.

Goffman, Erving (1959): The Presentation of Self in Everyday Life. Middlesex: Penguin Books Edition (1969).

Haddon, Leslie (ed.) (1998): Communications on the Move: The Experience of Mobile Telephony in the 1990s, COST248 Report, Telia AB, Fasta. URL: http://www.cost269.org/.

Haddon, Leslie (2003): Domestication and Mobile Telephony. In: Katz, James E. (ed.): Machines that Become Us: The Social Context of Personal Communication Technology. New Brunswick, New Jersey: Transaction Publishers, 43-56.

Haddon, Leslie / Vincent, Jane (2009): Children's Broadening Use of Mobile Phones. In: Goggin, Gerard / Hjorth, Larissa: (eds.): Mobile Technologies from Telecommunications to Media. New York: Routledge, 37-49.

Hartmann, Maren / Berker, Thomas / Punie, Yves / Ward, Katie (eds.) (2005): Domestication of Media and Technology. Maidenhead: Open University Press.

Hochschild, Arlie R. (1983): The Managed Heart Commercialization of the Human Feeling. Berkley: University of California (20th edition with afterword 2003).

Hochschild, Arlie R. (2003): The Commercialization of Feelings: notes from home and work. Berkeley: University of California.

Hochschild, Arlie R. (2008): Feeling Around the World. In: Contexts, Spring Vol 7 (2), 80. URL: http://contexts.org/articles/spring-2008/feeling-around-the-world/.

Höflich, Joachim R. (2003): Mobile Communication and the Situational Arrangement of Communication Behaviour. In: Nyíri, Kristóf (ed.):

Mobile Democracy: Essays on Society, Self and Politics. Wien: Passagen Verlag, 33-54.

Höflich, Joachim R. (2009): Mobile Phone Calls and Emotional Stress. In: Vincent, Jane / Fortunati, Leopoldina (eds.): Electronic Emotion: The mediation of emotion via information and communications technologies. Oxford: Peter Lang (UK), 63-83.

Katz, James E. (ed.) (2003): Machines That Become Us, The Social Context Of Personal Communication Technology. New Brunswick: Transaction.

Ling, Richard (1998): "It's OK to be Available": The Use of Traditional and Mobile Telephony amongst Norwegian Youth. Paper presented to the XIV World Congress of Sociology, 'Social Knowledge: Heritage, Challenges, Prospects', Montreal, 26th July - 1st August.

Ling, Richard (2008): New Tech New Ties How mobile communication is reshaping social cohesion. Cambridge: MIT.

McLuhan, Marshall (1964): Understanding Media: The Extensions of Man. NewYork: McGraw Hill.

Meyrowitz, Joshua (1985): No Sense of Place; the impact of electronic media on social behaviour. New York: Oxford University Press.

Maldonado, Tomás (2003): The Body: Artificialization and Transparency. In: Fortunati, Leopoldina / Katz, James E. / Riccini, Raimonda (eds.) (2003): Mediating the Human Body: Technology, Communication and Fashion. Erlbaum: Mahwah, New Jersey, 15-22.

Mobile Data Association (2009): The Q4 2008 Mobile Trends Report Press Release 13 February 2009. URL: http://www.themda.org/mda-press-releases/the-q4-2008-uk-mobile-trends-report.php (15 April 2009).

Ofcom (2008): The Communications Market 2008. URL: http://www2.ofcom.org.uk/research/cm/cmr08/ (15 April 2009).

Pertierra, Raul (2005a): Enabling Technologies: Extending the Senses Beyond Body-to-body Communication – Texting in Online Relationships in the Philippines. Paper presented at the International Conference on Mobile Communication and Asian Modernities II, Beijing, 20th-21st October.

Pertierra, Raul (2005b): Mobile Phones, Identity and Discursive Intimacy Human Technolog. In: An Interdisciplinary Journal on Humans in ICT Environments 1(1), 23-44. URL: www.humantechnology.jyu.fi.

Silverstone, Roger / Haddon, Leslie (1996): Design and the Domestication of Information and Communication Technologies: Technical Change and Everyday Life. In: Silverstone, Roger / Mansell, Robin (eds.): Communication by Design. The Politics of Information and Communication Technologies. Oxford: Oxford University Press, 44-74.

Short, John / Williams, Ederyn / Christie, Bruce (1976): The Social Psychology of Telecommunications. London: Wiley.

Tafazolli, Rahim (ed.) (2006): Technologies for the Wireless Future. In: Wireless World Research Forum (WWRF) Vol 2. London: Wiley.

Vincent, Jane (2004): 11-16 Mobile. Examining Mobile Phone and ICT Use Amongst Children aged 11-16. URL: www.dwrc.surrey.ac.uk (January 2008).

Vincent, Jane (2006): Emotional Attachment and Mobile Phones. In: Knowledge Technology and Policy (19), 29–44.

Vincent, Jane (2007): Emotion and my mobile phone. In: Conference Proceedings Towards a Philosophy of Telecommunications Convergence, 327-332. Conference organised by T-Mobile Hungary and The Hungarian Academy of Sciences Budapest.

Vincent, Jane (2009a): Emotion, my mobile, my identity. In: Vincent, Jane / Fortunati, Leopoldina (eds.): Electronic Emotion: The mediation of emotion via information and communications technologies. Oxford: Peter Lang (UK).

Vincent, Jane (2009b): Me and My Mobile. In: Fortunati, Leopoldina / Gebhardt, Julian / Vincent, Jane / Petrovčič, Andraž / Vershinskaya, Olga (eds.): Interaction in Broadband Society. Berlin: Peter Lang, 187-206.

Vincent, Jane / Fortunati, Leopoldina (eds.) (2009c): Electronic Emotion: The mediation of emotion via information and communications technologies. Oxford: Peter Lang (UK).

Vincent, Jane / Harper, Richard (2003): Social Shaping of UMTS – Preparing the 3G Customer. In: Report 26 UMTS Forum. URL: http://www.umts-forum.org accessed via www.dwrc.surrey.ac.uk (April 2008).

Vincent, Jane / Haddon, Leslie (2003): Informing Suppliers about User Behaviours to better prepare them for their 3G / UMTS. In: Customers UMTS Forum Report No. 34. URL: www.dwrc.surrey.ac.uk (May 2008).

Fashion and the mobile phone: a study of symbolic meanings of mobile phone for college-age young people across cultures

Satomi Sugiyama

Introduction

During the past years, numerous scholars have examined the social interactions involving the mobile phone in public places. One of the perspectives taken for the examination is to situate the mobile phone as a symbolic object. Domestication Theory (Silverstone / Haddon 1996) as well as Apparatgeist Theory (Katz / Aakhus 2002) draws our attention to the symbolic meanings of the mobile phone. Domestication Theory posits that media and information and communication technologies are "not just material or functional objects but have a powerful symbolic charge" (Silverstone / Haddon 1996: 59 et seq.). Suggesting the importance of taking the social interaction perspective in understanding the symbolic meanings of technologies, the theory holds that the "symbolic charge is itself the product of the activities of those, together, design, market and use technologies" (ibid.: 60). In line with Domestication Theory, Apparatgeist Theory (Katz / Aakhus 2002) posits that users, non-users, and anti-users of technology interact with one another, and collectively negotiate meanings of the mobile phone. Since the theory treats the technology as analogous to humans, the interactions occur not only between users (as well as nonusers and rejecters) of technologies but also between human and technologies as well as technologies themselves. These interactions construct aesthetic meanings for the mobile phone since "[t]he technology itself offers people an opportunity to modify preconceived uses, and consequently, the way in which its design and vector develop relies on these modifications" (ibid.: 301).

In addition, the theory proposes that there is a relationship between mobile technologies and self-expression. That is, it suggests that examining how the body and the way people dress their bodies using technologies informs people's adoption decision and mobile phone behavior. Katz and Aakhus argue that people oscillate between explicit reasons and implicit reasons for using the mobile phone. Explicit reasons refer to form, function, and price; implicit reasons refer to how others perceive the mobile phone users, beliefs about the usefulness, and appropriateness to one's concept of "self" (ibid.: 309 et seq.). Apparatgeist Theory suggests that it is important to consider the sense of self and its expression in the social interactions involving mobile communication technologies. That is, the theory draws attention to the importance of considering the "sense of self" of the individuals who use or do not use the mobile phone, as well as the "sense of self" of those who use the mobile phone in variant ways.

Stemming from the Apparatgeist Theory, but focusing more on the relationship between technology and the human body, Katz (2003) proposes a *Machines That Become Us* perspective. The term *Machines That Become Us* implies that "the technologies 'become' extensions and representative of the communicator," "technologies become physically integrated with the user's clothing and even body," and technologies are "becoming to the wearer" (Katz 2003: 1). Katz conceives of media technology to include telephone wires, computer networks and mobile phones, "second skin", fashion, accessories, and decorations that we wear on our body (ibid.: 18). It claims, in conjunction with the Machines That Become Us perspective, that the technology does not only extend human capacity in a rather direct way (e.g. phone-ear and voice, wheel-foot), but that it is also being *incorporated* into the human body and becoming part of a means of expressing identity and emotions. In addition, it explicitly extends the research focus to include such notions as fashion and the body, as well as how the mobile phone relates to the body and the sense of self. All of these theoretical approaches lead us to examine symbolic meanings, including self-expression, the design aspect of the instrument, and the image of the technology.

In order to examine the symbolic meanings of the mobile phone further, we turn to the concept of *fashion* and its theories. Fashion seems to be a rather overlooked concept in the study of interpersonal communication.[1] Most research that directly examines the notion of fashion in the field of communication proceeds from the perspective of marketing and advertising research (e.g. Miller / McIntyre / Mantrala 1993; Englis / Solomon / Ashmore 1994; O'Cass 2004; Banister / Hogg 2004) or from a cultural studies perspective (e.g. Skov 1996; Negrin 1999; Wilson 2000). Not only in the study of interpersonal communication, but also in academe in general, fashion does not seem to be perceived as a worthy topic for serious scholarly research (Entwistle / Wilson 2001; Tseëlon 2001). However, the notion of fashion offers guidance into a way of understanding appearance, and in turn, into understanding the communicative expression of the self. Some scholars noted the importance of the concept and applied it to examine mobile communication (e.g. Campbell 2008; Fortunati

1 Fifty-nine articles were found with the key word "fashion" on the database of Communication and Mass Media Complete. Thirty-eight of the articles were published in marketing or advertising journals. Only a few articles (book reviews) on fashion were found in such journals as the Journal of Communication and the European Journal of Communication (searched on June 2, 2005).

2002, 2003a, 2003b; Fortunati / Cianchi 2006; Fortunati / Katz / Riccini 2003; Katz / Sugiyama 2005, 2006; Ling 2003). The research explicitly guided by the fashion theory, however, is still limited.

The concept of fashion is conceived of as a clothing-related term, along with such terms as adornment, appearance, dress, and style, terms that are sometimes confused with one another (Kaiser 1997: 3 et seq.). Sproles (1979) states that fashion is a "dynamic social process by which new styles are created, introduced to a consuming public, and popularly accepted by that public" (ibid.: 5). According to Lowe and Lowe (1985), fashion involves "rapid, incessant change, spread by social contagion where the emphasis is on novelty, not enhanced utility" (ibid.: 204-205). Crane (2000) posits that fashion entails "strong norms about appropriate appearances at a particular point in time" (ibid.: 1). Davis (1992) holds that the element of change is essential in its definition (ibid.: 14). Taking these together, fashion can be defined as a dynamic and rapid social process of changing norms about appropriate appearances at a particular point in time.

While fashion is a social process, clothing is "any tangible or material object connected to the human body" and appearance is "the total, composite image created by the human body and any modifications embellishments, or coverings of the body that are visually perceived; a visual context that includes clothing as well as the body" (Kaiser 1997: 4). As such, the mobile phone in so far as it tends to follow wherever one goes, and is closely connected to the human body, could be considered as a part of clothing, contributing to one's overall appearance.

With these conceptualizations of fashion, clothing, and appearance, the notion of fashion serves as a social framework for understanding communicative behavior via one's appearance involving clothing and the body. One important aspect of fashion is its intersubjective nature. As Kaiser (2001) posits:

> "In many ways, 'subjectivity' (a way of being and becoming in the world) is to intersubjectivity (collective understandings of 'how things are') what style is to fashion, and truth is to knowledge. In each case, the former term suggests individual actions and intentions, whereas the latter implies collective negotiations and understandings" (Kaiser 2001: 80).

This intersubjective nature of fashion is further complicated by the intertwining of individual and collective actions (Kaiser 2001). The stylistic diversity and complex intersubjectivity that fashion entails, then, creates ambivalent feelings about one's appearance. In turn, examining fashion and appearance as they relate to the mobile phone, promises to yield rich symbolic meanings for the mobile.

Aforementioned theories on technologies as well as fashion also emphasize that it is important to understand the symbolic meanings of the mobile phone within a cultural context. As Höflich (2003) argues, it is pivotal to maintain orientation in social space as well as orientation toward others in that space. In line with this position and drawing on the case of Chinese culture, Yu and Tng (2003) also argue for the importance of paying attention to culture-specific meanings of the mobile phone. The following statement by Kaiser (2003) further elucidates the relevance and importance of culture in examining symbolic meanings of the mobile:

> "Inasmuch as fashion and style represent entangled freedoms and ambiguous subjectivities, they offer heuristic strategies for thinking through struggles and pleasures that might otherwise not be expressed, if we take the time and energy to unravel their potential in the context of a frenetic, global technoculture" (Kaiser 2003: 160).

Focusing on the notion of fashion, this paper analyzes how college-age young people across cultures perceive the mobile phone as an aestheticized object that expresses and defines the self. Employing the concepts of fashion, clothing, and appearance, the paper seeks to understand the cultural meanings of the mobile phone appearance by highlighting convergence and divergence across cultures.

Methods

Focus group interviews and individual interviews were conducted during 2005-2008 in various cultural settings. The data from Japan consists of four sessions of focus group interviews conducted early in 2005. Japanese college students were recruited for the study at a private university in Nagoya. Thirty students participated (6 males and 24 females). For the U.S. data, three focus group interviews were conducted at a private Liberal-Arts university in the Northeast in 2006. Twenty students participated in total (3 males, 17 females). To further diversify the study, two sessions of focus group interviews and 17 individual interviews were conducted at a private Liberal Arts college in Switzerland during 2007-2008. The focus group interviews involved 11 students (4 males, 7 females) from various cultural backgrounds (e.g. 6 from the U.S., 2 from Jordan, 1 from Syria, 1 from Russia, 1 from Honduras). Participants of individual interviews also included students from various cultural backgrounds: 4 from the U.S., 10 from the Middle East, 1 from Russia, 1 from Africa, and 1 from South America. These individual interviews took a slightly different interview schedule, but they compliment the focus group interview data by broadening the cultural perspectives. All interview participants were considered as appropriate in that they have "appropriate experience in the cultural scene", and "the ability and willingness to articulate his or her experience in the interview context" (Lindlof 1995: 178). Each interview session lasted 1 to 1.5 hours, taking the semi-directive approach. All interviews were audio-recorded, and transcribed.

Findings

Convergence: the mobile phone as an aestheticized object

One key theme that emerged from the interview data across cultures is the perception of the mobile phone as an aestheticized object. Although to varying degrees, the interview participants indicated they actively evaluate the style of the mobile phone including such aspects as stylistic features and color. This aspect became apparent in examining the reason why they chose their current mobile phone and the reason why they changed their mobile for a new one. For example, the Japanese participants commented:

> "I just use (my phone) to make phone calls and exchange mails (i.e., SMS and MMS), so I would rather have a cute phone with ease of use" (Japan FGI 3, female).

> "I choose (my phone) thinking that it's fine as long as the phone is for DoCoMo, but I had a hard time at the beginning since I have to get used to the new phone (...) I just want to make phone calls and exchange mails, so it's even OK not to have a camera feature, and it's really OK if I can use mails and telephone, so its look (is more important) than its features...the one that feels right" (Japan FGI 2, female).

> "I actually don't like the clam-shell type. I like a really straight shape, but it's not available any more. So I have to compromise in this regard" (Japan FGI 3, male).

Not only the importance of the stylistic form, but some also showed rather strong preference about the color of their mobile, wanting black, red, orange, and so forth. For them, color was not something that they chose because there happened to be some options. They wanted black, red, or orange. In the case of the participant who wanted an orange phone (Japan FGI 4), she needed to compromise by choosing a phone with the closest in color.

It is interesting to note that no one in the Japanese focus groups mentioned the issues that relate to the most basic function of the mobile phone, such as its battery life and good reception, as a consideration for selecting the particular phone. For the focus group participants, these basic but essential aspects for the mobile phone to function as a telephone in transit do not appear to be primary considerations in selecting their mobile phone at this time, perhaps due to the already reasonably good reception and battery life available to them. Nonetheless, they did mention that their battery usually dies in a year or so. Rather than trying to look for a phone with a longer battery life, however, their attention seems to be directed to the aesthetic aspects of the mobile, so much that they described their having to settle for a style and color as a compromise. The mobile phone has to be "easy to use", which certainly indicates that the mobile phone is a functional object for them. The mobile phone has to be within their budget as well, which is a reasonable consideration, given their status as college students. Nonetheless, style, color, new features, and newness were also found to be major criteria in selecting a mobile phone. Considering new features as a contributing stylistic aspect that goes beyond the originally intended function of the mobile phone as a telephone in transit, these criteria that emerged from the focus group interviews suggest the importance of a style aspect of the mobile phone. In addition to the importance of the stylistic form of the mobile phone, the importance given to the color indicates how the mobile phone has become an aestheticized object whose stylistic evaluation is based on one's subjective judgment.

Although to a different extent, some of the U.S. focus group participants also expressed the importance of the style of the mobile phone. For example, a participant in the U.S. FGI 3 expressed that she chose her current phone because it looked "pretty". Another participant from the U.S. FGI 1 stated that she would like the "SLVR" from Motorola next time she changes her phone, which was a new phone at the time of the interview. Furthermore, the focus group participants of various cultural backgrounds in Switzerland also indicated the importance of the mobile phone look in selecting their mobile as illustrated below:

> "Like the way it looked - I thought it looked good" (Russia, F).
> "Me too, I like the style" (Syria, F).
> "Just liked the appearance" (Greek-US, F).
> "thinnest phone in the market" (US, M).

As the above data illustrate, the mobile phone is not a mere telephone medium that one uses to communicate at a distance in transit. Although the mobile phone is expected to serve for its originally intended function and desired to be easy to use, it is also becoming an object that one chooses based on one's subjective taste among available options. It appears to resemble clothing and personal accessories such as bags and shoes. It has to be able to bind spaces that hinder communication, in the same way that clothing has to protect one's body from the wind, shoes have to be able to protect one's feet from gravels, and a bag has to hold personal items. At the same time, for many users, the mobile phone also has to fit to their aestheticism, in the same way that clothing, shoes and handbags have to for many wearers. An Arab female participant's comment supports this point: "Phone is like bags and shoes. Other people buy a lot of shoes, I buy cell phone". Although it is certainly important to acknowledge variant degrees to which one sees the mobile phone as an aestheticized object, the data certainly suggest that these participants engage in the aesthetic evaluation of the mobile based on their subjective aesthetic standard.

Convergence: emergence of the mobile phone fashion

The previous section discussed the way that the mobile phone style is evaluated based on personal taste and subjective judgment. A notable point here is that the style evaluation appears to change over time. As a result of the change, the phone that used to fit their taste ends up becoming an object that they "get tired of" (Japan FGI 1, 2 & 4), contributing to the reason why they start wanting a new mobile phone. This "change" in stylistic evaluation signals the impact of time. Over a year or so, the participants develop a feeling of "tired of" for the phone that they carefully selected at the time of purchase. This impact of time seems to be an underlying factor for the other major reasons that emerged, which were "wanting new and improved features", "wanting a phone that looks like others", and "wanting the newest phone". These reasons point out the concerns that the focus group participants have; that is, how does my phone look like in relation to others' phones?

The stylistic evaluation of one's own phone in relation to others' phones was apparent in wanting a phone like others. The following quote (Japan FGI 2) illustrates this point:

> "(...)
> Male 1: In addition to bad function, I wanted a clam-shell type. I also wanted the one with a camera feature.
> Moderator: Why did you want a clam-shell type?
> Male 1: Fashion.
> Moderator: (The type) became more a mainstream?
> Male 1: I thought that a non-clam-shell type was out of date.
> Moderator: What about others?
> Male 3: I chose this phone because features of the mobile phone have improved as well.
> Moderator: "As well"?
> Male 3: Well, other people started changing into nice ones, and I thought mine looked so old! (...)"

This quote suggests that these male students were concerned not only with the design, such as additional and improved features of the mobile phone, but also a great deal with its style. And the design and style of their mobile phone were evaluated based on what others have. And as the terms "out of date" and "so old" suggest, the design and style of "what others have" and those of "what I have" appear to be evaluated in relation to what are perceived as new or old: the norms of mobile phone style and design at the moment. In this way, time becomes an essential factor in their aesthetic evaluation.

As will be discussed later in the paper, the U.S. focus group participants did not seem to emphasize the importance of the mobile phone being one of the newest. However, they did express their desire for their mobile phone to conform to the current "standard" or "normal" look. For example, a female participant from the U.S. FGI 2 stated:

> "I already had another Verizon phone that was a more obsolete model, and we wanted to get an upgrade, which included a new model of the phone. So this is just the *standard* new model, and a lot of my friends have this phone because it is kind of *standard*, most simple phone, meaning it doesn't have a camera or any of those extra features, but it's still pretty advanced because it flips open and has the little screen."

Another participant of the U.S. FGI 2 commented:

> "Mine I got when it was time for my first upgrade after I had my first phone for two years, so I got one that had a camera phone because I thought that was kind of cool-- I don't really use it that much, but I just got whatever the normal one with a camera was."

Although they do not emphasize the importance of the mobile phone style, their comments indicate that there is an understanding of the standard or normal look for the mobile.

Not only is time an important part of the style and design evaluation of the mobile phone, but also the people who carry the mobile phone appear to become part of the evaluation of the mobile phone style / design. As such, the mobile phone serves as an embellishment of the human body, contributing to the composite image of the person in relation to others. That is, the evaluation of the mobile phone style / design does not exist in a vacuum. The style / design of the mobile phone is evaluated in conjunction with its users, that is, its *wearers*. Thus, it becomes apparent that the way the mobile phone style / design is evaluated is largely based on the mobile phone's *appearance*: the total composite image that surrounds the mobile phone. The mobile phone is not only an object whose *style* is evaluated based on one's subjective standard, but also an object whose *appearance* is evaluated based on the collective understandings of how the mobile phone should look at a given point in time. Through this social interaction that involves the mobile phone, subjective style / design evaluations become intersubjective. Thus the mobile phone *fashion* emerges.

The process of creating a collective understanding of how the mobile phone should look like is exemplified in the conversation that took place during a Japan focus group interview about "Infobar." The model called "Infobar" came up for discussion during the two interview sessions. The "Infobar" is a model that debuted in October 2003, about a year and a half before the time of the interview. It placed an emphasis on design rather than function. Many of the participants recognized the model although nobody actually owned it, suggesting that they have either seen its advertisement, news report, store display, or learned about it from people around them. A notable point in the discussion was that they were aware of who among their friends or classmates had the Infobar phone. One reported that some of her friends bought it right after it came out. In sharing these episodes, the participants' evaluations of the model were revealed: some thought that the "infobar" was cool (*kakkoii*) or cute (*kawaii*), and others thought that it looked like a toy. One said that she wanted it because of its straight shape, looking very different from the mainstream clamshell type which everybody has.

A similar conversation took place during a mixed-culture focus group interview. The participants discussed how some of their friends have Dolce and Gabbana's gold Razr phone, and they shared their subjective evaluations as "too much" (U.S.), "kind of cool" (Russia), and "I don't like it" (Syria).

These comments show how they actively engage in the process of style evaluation through the everyday observation of the mobile phone that others carry. The data illustrate that these subjective evaluations are intersubjectified, contributing to the constantly changing norms of mobile phone appearance.

It appears that there is a rapid social process of changing norms about appropriate appearance of the mobile phone at a particular point in time. By looking at others' mobile phone, and new phones that become available in the market, certain stylistic evaluations become intersubjectified among some group members, yielding a sense of what are appropriate and inappropriate norms of mobile phone appearance. And the intersubjectified stylistic evaluations serve as frameworks for evaluating one's own mobile phone, in conjunction with one's subjective aesthetic judgment. The mobile phone is an object whose style is evaluated within a rapid social process of changing norms about appropriate appearance at a particular point in time. The process of mobile phone fashion appears to involve dynamic social interactions among users, observers, and the industry, just as Apparatgeist Theory suggests.

Divergence: the mobile decoration

Although the participants of the study perceived the mobile phone as an aestheticized object and certain degree of the mobile fashion exists across the cultural contexts examined, the extent to which the participants from various cultures perceive the mobile phone as an aestheticized object seems to be different and requires a nuanced analysis. A point that particularly stands out is the mobile phone decoration observed in the case of Japanese participants. As the table 1 shows (see Appendix), the Japanese focus group participants (total of 30 participants) had various types of decoration on their mobile on the day of the interview. Among the 48 non-Japanese focus group / individual interview participants, however, only 3 participants had some type of decoration. This makes a stark contrast between the Japanese participants and non-Japanese participants. How can we make sense of this contrast?

A female participant from the U.S. (FGI 1) decorated her mobile with "stick-on jewels" to "jazz up" her "ugly" phone. Another female participant from the U.S, one who participated in the mixed-cultural focus group interview, had two small charms, which she explained that she just thought, "they were cute" (mixed FGI 2). Another female interview participant from Bahrain had a small charm of a football. She explained that it was a charm of the Manchester United football team, and that it was a gift from her friend who knows that she is a fan of the team. She attached the charm to her mobile because "nobody has this in Bahrain". These reasons are in line with a key explanation that emerged from the Japanese focus group interviews; that is, expressing their individuality via the mobile phone appearance. For example, the following exchanges took place during the focus group 2:

> "(…)
> Female 1: A mobile phone is something that others also have. So, (I decorate mine) to mark that this is mine.
> Moderator: Do you mean that you can tell that it is yours?
> Female 1: Well…when I think "oh that person also has the same phone," because of what is attached (e.g. an accessory), I can tell the one I have is mine…
> Male 1: (And you do this) to show individuality.
> Moderator: For individuality?
> Female 1: Yes.
> Moderator: Do others share the same feeling? (That you attach things) to show individuality?
> Female 2: If I don't put anything…well…it's the same as others… (I do it) so that I can tell more or less…I'm not thinking much…
> Female 3: I don't think much either…
> Male 2: I have that sense of showing individuality (with the decoration), but nowadays, I feel that not putting anything (on my mobile phone) is my own individuality, so I don't put anything. (…)"

Although two in the above conversation stated "I'm not thinking much", it is interesting to note that there is a basic sentiment among the participants that they seek to mark their mobile phone as their own: to make it different from what others have. A similar sentiment was found in other focus groups. For example, three female students, who are close friends with each other, expressed their desire to avoid buying the exact same phone (Japan FGI 4). The three female students owned the same model, but they each had red, black, and white variations in color. They said that they happened to choose the same model, and did not set out to buy matching ones. The first person that purchased the phone among the three chose red. The second person said she decided on the black one because she was going to purchase a same model as her friend's. The third person denied her intentional choice of white color to avoid the colors that her friends already had, but it is interesting that she chose the last available color of the model that did not match with her friends.

The example above illustrates how Japanese participants seek to differentiate their mobile phones. For them, the mobile phone, an aestheticized and expressive object, like clothing, is a personal front (Goffman 1959) that allows them to present their individuality. It is an object that symbolically expresses who they are. Then, it is not surprising that numerous participants explained that some of their mobile phone decorations included their favorite characters (e.g. Hello Kitty, Snoopy, and Disney characters) and personal taste (e.g. silver charms).

Expressing individuality, however, was not the only reason given to explain their decorations. What emerged from the Japanese focus group interviews is that those charms, stickers, and photo-stickers are largely understood as relational symbols. For example, two of the participants in the Japanese focus group 4 had a matching charm of a "duck," signaling their friendship. Many of them had *purikura* stickers on the surface of their mobile, showing with whom they are friends. These *purikura* stickers are understood as a social currency (Chalfen / Murui 2004), and by placing them on the surface of the mobile, these Japanese participants sought to manifest that they are unique individuals tied to other unique individuals (Sugiyama 2009).

As described above, many Japanese participants pay considerable attention to the way their mobile phones express the self and social relationships. Furthermore, this expression of the self and social relationships appears to be carefully managed in relation to their understanding of who they are not. The act of separating themselves from others in terms of the appearance of the mobile phone was observable in numerous occasions during the focus group interview. They discussed the way they associate a certain style, use, and display of the mobile phone with a particular social group.[2] For example, the following conversation (Japan FGI 4) illustrates:

"(…)
Female 1: High school students, they all put big ones!
Moderator: They put big ones?
Female 1: Yeah, big ones.
Female 2: Yeah, bigger than the mobile phone itself.
Female 1: Yeah, (it was) bigger!
Female 3: I did not put too many things even when I was in high school.
Female 2: Some of those (decorative items) were really something.
Female 4: Yeah, like "what's that?!"
Female 5: It was more like a doll than a mobile phone!
All: (laugh)
Female 4: It was like a bag can get all filled up with the phone (with the decorative items) alone! (…)"

The way they talked about current high school students whom they observe in their daily life, as well as the way they talked about themselves when they were at high school, signals that they clearly separate themselves from this "younger"

---

2 Ling (2003) discusses how certain mobile phone use and display styles are talked about as *harry* (tacky / vulgar) or *soss* (preppy / pompous) by Norwegian teens. His discussion makes a good comparison with the case of Japanese young people, highlighting a similarity across cultures.

group of people. Some of them used to be a part of the "crude" mobile phone appearance of high school / junior high school students, but they are not any more. The way that they talked about high school and junior high school students in a teasing manner suggests that they see their mobile phone appearance as much more sophisticated as an appearance, which fits that of some grown-ups like themselves.

Another social group frequently mentioned were people who generally fall under a category of middle-aged men (*ojisan*) and businessmen (*sarariiman*). The following exchanges of the Japanese focus group 1 illustrate the way they dissociate their mobile phone appearance from that of middle-aged men:

"(...)
Moderator: Do you notice how others carry their mobile phone in public places?
Female 1: Some hang the mobile phone around their neck...
Female 2: Yeah, there are many (who hang their mobile phone around their neck).
Female 3: Yeah, there are those.
Female 4: Many ojisan do it.
All: (laugh)
Female 5: Ojisan put the mobile phone in their chest pocket.
All: Yeah!
Moderator: So, that's the ojisan style?
Female5: Yeah, ojisan. Sarariiman wearing suits puts the phone in the inside pocket of his jacket.
Female 6: Some use a clip.
Several simultaneously: Yeah!
Moderator: That's ojisan style?
Female 7: But there are many who put their mobile phone in a case.
All: Yeah! I know what you mean!
Female 7: I was surprised. They put the mobile phone in a case!
All: (laugh)
[...]
Moderator: So your male friends do not keep their phone in this pocket? (pointing to the chest pocket)
All: (laugh, and simultaneously) I've never seen it! / Oh, no! (...)"

This conversation indicates that these first-year female students clearly differentiate the appearance of mobile phones owned by men of their "in-group" from men of their "out-group." Such appearances as "hanging around neck," "keeping in a chest pocket," and "using mobile phone case" are all attributed to the appearance of middle-aged men from the perspective of these female college students. There appeared to be a general consensus in this regard at least among the participants of the focus group 1. When the moderator asked if their male friends carry the mobile phone in similar ways, their primary reaction was a laugh, complemented by such remarks as "Oh, no!" and "I've never seen it!" This suggests that the mobile phone appearance that is associated with the middle-aged men does not seem to fit that of the male members of their "in-group" to the point that they find the suggestion of similarity to be almost ridiculous. It seems that the mobile phone appearance observed among middle-aged men, from the perspective of these female college students, is something that they do not quite understand. Or rather, it is something that does not quite satisfy their sense of what fits them or their in-group members. Thus, the mobile phone can be said to be an aestheticized object that can *become* to certain group of people if it is appropriately incorporated into one's appearance, but the same mobile phone appearance might not *become* to other group of people.

On the other hand, as we saw earlier in the data of the U.S. focus group interviews, U.S. focus group participants reported their contentment with the look of the mobile phone as long as they conform to the "standard" look; that is, their mobile phone needs to be just like others' phone. This also can be interpreted to mean that their mobile phone *can* look just like others' phone. Furthermore, what is noticeable here in the U.S. case is that the term "others" includes the vast majority of mobile phone users and does not segment users into sub-cultural groups such as business people and high school students in the way that Japanese participants did. The following exchanges took place during the U.S. focus group 1:

> "(...)
> Female 1: It was one of the cheaper ones... I think all I knew was that I wanted it to have the screen so you could see who was calling. So I think it was the cheapest one with the screen.
> Female 2: Yeah, same – I wanted the screen, and also the deal was "Buy one and get one free." So my brother and I just did it together. But I didn't really care about the camera or anything. I just wanted a flip phone with the screen.
> Female 3: Same...my mom and I got the same one because it was "Buy one, get one free." (...)"

They did not indicate their desire for distinguishing their phone from their father's phone, their mother's phone, or their younger sibling's phone. That is, their mobile phone appearance is not reserved only to themselves. All of their family members can have a silver tiny "flip" phone with the logo of "verizon", for example, and that is the way it is. If their understanding of the mobile phone were similar to that of the Japanese participants, they would have indicated their discontent for matching phones with their family members. They would also have decorated their phones with some charms and stickers to indicate their subjective tastes as well as intersubjective aesthetic values proper to their particular social group. Nonetheless, they seemed to be "happy enough" to have a reasonably functioning mobile with a standard look – standard for everyone, including their younger siblings and older people such as their mother and father.

Conclusion

The present study showed that, for the interview participants across cultures, the mobile phone is clearly an aestheticized and symbolic object subject to social interactions that create the norms for the mobile phone appearance. These young people are not necessarily aware of the emergence of the mobile phone fashion, since this phenomenon falls under the third level of the social consequences that Katz (1999) described. The mobile phone is becoming a cultural icon. However, what serves as a cultural icon is not the phone per se, but the mobile phone appearance. As the present study showed, various types of mobile phone appearance are associated with certain sub-groups within a society, yielding symbolic meanings for the mobile phone appearance. The conceptualizations of fashion, clothing, and appearance, therefore, served as useful tools for understanding the symbolic aspect of the mobile phone by providing a lens to analyze what is observable on the surface of the mobile phone, leading to the critical consideration of culturally-situated meanings of the mobile.

The question of cultural differences and similarities as related to the symbolic aspect of the mobile has been a part of previous scholarly discussions. Many have already argued that the style of the mobile phone is very important for young people, since it expresses their identity. For example, Ling and Yttri (2002) argued the importance of the mobile phone style for Norwegian teens. Fortunati (2002) argued the importance of the style aspect of the mobile phone in Italy. Lobet-Maris (2003) posited that young people in Belgium see the style of the mobile phone as a crucial motivation for choosing their own mobile phone, and that young people invest their mobile phones with symbolic significance, serving as a large factor for constituting personal identity. Katz, Aakhus, Kim and Turner (2003) found notable consistencies across cultures that they have examined (the U.S., Korea, Norway, and Namibia) particularly in terms of the way the information and communication technologies related to the concepts of fashion and the body.

The present study largely offers the results in line with those previous studies, suggesting that some type of global symbolic meanings of the mobile phone might be emerging. However, the subtle differences within culture as well as across cultures should not be overlooked either. Green (2003) challenges the position that assumes "teenagers will consume if they are able to do so, in subcultural and 'global youth' styles" (ibid.: 213) by pointing out the significant diversity among youths. Green concluded that there is a significant diversity among young people in terms of the meaning and use of the mobile phone. She argued further that this diversity could be attributed to the contexts of their lives, depending:

> "not only on the age of the young people concerned, but also on the school and household contexts and their relative negotiation of identity, difference, and independence among themselves and with others" (Green 2003: 213).

Indeed, the U.S. participants in the U.S. focus groups indicated quite a different understanding of the mobile compared to the understanding of the U.S. participants in the mixed-cultural focus groups conducted in Switzerland. As the college-age young people originally from the U.S. interact with others from various cultural backgrounds in the European cultural context, they may have developed an understanding of the mobile quite different from that of those who grew up and studied in a relatively culturally homogenous environment. This calls for a careful analysis of cultural meanings as well as reconsideration of the cultural boundaries defined by the traditional national boundaries. The way people understand this interpersonal communication technology is not necessarily predicated on the traditional cultural values and communication styles defined by a national boundary. Instead, something else that differentiates individuals makes multiple segmentations within a national boundary.

At the same time, some traditional cultural characteristics defined by national boundaries might not be disappearing either. Focus group interviews of young Japanese people offer an opportunity for a more nuanced analysis of symbolic meanings in the context of Japan. Some styles, use, and decoration of mobile phones that were reported during the focus group interviews seem to be quite particular to Japanese culture. De Gournay and Smoreda (2005) emphasize the uniqueness of the way Japanese women use the mobile phone. Barry and Yu (2002) report that Japanese youths store both practical and emotional memories in the mobile phone in the form of information as well as stickers and mascots. These findings, which are consistent with what the present study found, led them to argue that "[t]he phone can be a repository for precious memories because it is completely trusted and in sync with Japanese culture" (ibid.: 168). Ito (2005) warns not to use Japan "as a template for a mobile future in other countries" (ibid.: 2). How unique is Japanese mobile phone culture? Alternatively, are some of these lines of argument an echo of traditional views of Japanese culture as very mystic and unique?

At this point, as the data suggested, there seem to be some overall convergence across cultures in terms of the basic understanding of the mobile phone, but also, there seem to be something quite particular about the mobile phone appearance and meanings associated to the mobile phone by the Japanese college-age youths. This finding is in line with what Höflich and Hartmann (2006) claim; that is, the near-universal meanings across cultures are observable at the surface level, but culture-specific meanings emerge with a closer look of everyday use of the mobile. Further careful analyses of culturally situated symbolic meanings of the mobile will help us answer these questions. In this endeavor, the theoretical framework of fashion will continue to be a promising approach as it allows us to examine the *technology with people* in social and cultural contexts that evolve through time.

References

Banister, Emma N. / Hogg, Margaret K. (2004): Negative symbolic consumption and consumers' drive for self-esteem. In: European Journal of Marketing, 38 (7), 850-169.

Barry, Micheal / Yu, LiAnne (2002): The uses and meaning of i-mode in Japan. In: Revista de estudios de juventud, 57, 139-150.

Campbell, Scott (2008): Mobile technology and the body: Apparatgeist, fashion, and function. In: Katz, James E. (ed.): Handbook of Mobile Communication Studies. Cambridge, MA: MIT, 153-164.

Chalfen, Richard / Murui, Mai (2004): Print club photography in Japan: Framing social relationships. In Edwards, Elizabeth / Hart, Janice (eds.): Photographs Objects Histories: On the Materiality of Images. New York: Rourtledge, 166-185.

Crane, Diana (2000): Fashion and Its Social Agendas: Class, Gender, and Identity in Clothing. Chicago: University of Chicago Press.

Davis, Fred (1992): Fashion, Culture, and Identity. Chicago: University of Chicago Press.

De Gournay, Chantal / Smoreda, Zbigniew (2005): Space bind: The social shaping of communication in five urban areas. In: Nyíri, Kristóf (ed.): A Sense of Place: The Global and the Local in Mobile Communication. Vienna: Passagen Verlag, 71-82.

Englis, Basil G. / Solomon, Michael R. / Ashmore, Richard D. (1994): Beauty before the eyes of beholders: The cultural encoding of beauty types in magazine. In: Journal of Advertising, 23 (2), 49-65.

Entwistle, Joanne / Wilson, Elizabeth (2001): Body Dressing. New York: Berg.

Fortunati, Leopoldina (2002): Italy: Stereotypes, True and False. In: Katz, James E. / Aakhus, Mark (eds.): Perpetual Contact. Cambridge: Cambridge University Press, 42-62.

Fortunati, Leopoldina (2003a): The human body: Natural and artificial technology. In: Katz, James E. (ed.): Machines That Become Us. New Brunswick, NJ: Transaction, 71-87.

Fortunati, Leopoldina (2003b): Real people, artificial bodies. In: Fortunati, Leopoldina / Katz, James E. / Riccini, Raimonda (eds.): Mediating the Human Body. Mahwah, NJ: Lawrence Erlbaum, 61-71.

Fortunati, Leopoldina / Cianchi, Amalia (2006): Fashion and technology in the presentation of the self. In: Höflich, Joachim R. / Hartmann, Maren (eds.):

Mobile Communication in Everyday Life: Ethnographic Views, Observations, and Reflections. Frank & Timme, 203-226.

Fortunati, Leopoldina / Katz, James E. / Riccini, Raimonda (2003): Conclusion: Bodies mediating the future. In: Fortunati, Leopoldina / Katz, James E. / Riccini, Raimonda (eds.): Mediating the Human Body. Mahwah, NJ: Lawrence Erlbaum, 215-219.

Goffman, Erving (1959): The Presentation of Self in Everyday Life. New York: Anchor.

Green, Nicola (2003): Outwardly Mobile: Young People and Mobile Technologies. In: Katz, James E. (ed.): Machines That Become Us. New Brunswick, NJ: Transaction, 201-217.

Haddon, Leslie (2003): Domestication and Mobile Telephony. In: Katz, James E. (ed.): Machines That Become Us. New Brunswick, NJ: Transaction, 43-55.

Höflich, Joachim R. (2003): Part of two frames: Mobile communication and the situation arrangement of communicative behaviour. In: Nyíri, Kristóf (ed.): Mobile Democracy: Essays on Society, Self and Politics. Vienna: Passagen Verlag, 33-51.

Höflich, Joachim R. / Hartmann, Maren (2006): Mobile Communication in Everyday Life: Ethnographic Views, Observations, and Reflections. Frank & Timme.

Ito, Mizuko (2005): Introduction. In: Ito, Mizuko, Okabe, Daisuke / Matsuda, Misa (eds.): Personal, Portable, Pedestrian: Mobile Phones in Japanese Life. Cambridge, MA: MIT Press, 1-16.

Kaiser, Susan (1997): The Social Psychology of Clothing (2nd ed.). New York: Macmillan.

Kaiser, Susan (2001): Minding appearances: Style, truth, and subjectivity. In: Entwistle, Joanne / Wilson, Elizabeth (eds.): Body Dressing. New York: Berg, 79-102.

Kaiser, Susan (2003): Fashion, media, and cultural anxiety: Visual representations of childhood. In Fortunati, Leopoldina / Katz, James E. / Riccini, Raimonda (eds.): Mediating the Human Body. Mahwah, NJ: Lawrence Erlbaum, 155-161.

Katz, James E. (1999): Connections: Social and Cultural Studies of the Telephone in American Life. New Brunswick, NJ: Transaction Publishers.

Katz, James E. (2003): Machines That Become Us. New Brunswick, NJ: Transaction.

Katz, James E. / Aakhus, Mark (2002). Perpetual Contact. Cambridge: Cambridge University Press.

Katz, James E. / Aakhus, Mark / Kim, Hyo Dong / Turner, Martha (2003): Cross-Cultural Comparison of ICTs. In: Fortunati, Leopoldina / Katz,

James E. / Riccini, Raimonda (eds.): Mediating the Human Body. Mahwah, NJ: Lawrence Erlbaum, 75-86.

Katz, James E. / Sugiyama, Satomi (2005): Mobile phones as fashion statements: The co-creation of mobile communication's public meaning. In: Ling, Richard / Pedersen, Per E. (eds.): Mobile communications: Re-negotiation of the social sphere. London: Springer, 63-81.

Katz, James E. / Sugiyama, Satomi (2006): Mobile phones as fashion statements: Evidence from student surveys in the US and Japan. In: New Media and Society, 8(2), 367-383.

Lindlof, Thomas R. (1995): Qualitative Communication Research Methods. Thousand Oaks, CA: Sage.

Ling, Richard (2003): Fashion and vulgarity in the adoption of the mobile telephone among teens in Norway. In Fortunati, Leopoldina / Katz, James E. / Riccini, Raimonda (eds.): Mediating the Human Body. Mahwah, NJ: Lawrence Erlbaum, 93-102.

Ling, Richard / Yttri, Birgitte (2002): Hyper-Coordination via Mobile Phones in Norway. In Katz, James E. / Aakhus, Mark (eds.): Perpetual Contact. Cambridge: Cambridge University Press, 139-169.

Lobet-Maris, Claire (2003): Mobile phone tribes: Youth and social identity. In Fortunati, Leopoldina / Katz, James E. / Riccini, Raimonda (eds.): Mediating the Human Body. Mahwah, NJ: Lawrence Erlbaum, 87-92.

Lowe, Elizabeth D. / Lowe, John W. G. (1985): Quantitative analysis of women's dress. In Solomon, Michael (ed.): The psychology of fashion. Lexington, MA: Lexington Books, 193-206.

Miller, Christopher M. / McIntyre, Shelby H. / Mantrala, Murali K. (1993): Toward formalizing fashion theory. In: Journal of Marketing Research, 30(2), 142-158.

Negrin, Llewellyn (1999): The Self as Image: A Critical Appraisal of Postmodern Theories of Fashion. In: Theory, Culture & Society, 16(3), 99- 119.

O'Cass, Aron (2004): Fashion clothing consumption: antecedents and consequences of fashion clothing involvement. In: European Journal of Marketing, 38(7), 869-883.

Silverstone, Roger / Haddon, Leslie (1996): Design and the domestication of information and communication technologies: Technical change and everyday life. In: Mansell, Robin / Silverstone, Roger (eds.): Communication By Design. New York: Oxford University Press, 44-74.

Skov, Lise (1996): Fashion Trends, Japonisme and Postmodernism Or 'What is so Japanese about Comme des Garçons?' In: Theory, Culture & Society, 13(3), 129-152.

Sproles, George B. (1979): Fashion: Consumer behavior toward dress. Minneapolis, MN: Burgess.

Sproles, George B. (1985): Behavioral science theories of fashion. In Solomon, Michael (ed.): The psychology of fashion. Lexington, MA: Lexington Books, 55-70.
Sugiyama, Satomi (2009): Decorated mobile phone and emotional attachment for Japanese youths. In: Vincent, Jane / Fortunati Leopoldina (eds.): Electronic Emotion, the Mediation of Emotion via Information and Communication Technologies. Oxford: Peter Lang, 85-103.
Tseëlon, Efrat (2001): Fashion research and its discontents. In: Fashion Theory, 5(4), 435-452.
Wilson, Elizabeth (2000): A Response to Llewwllyn Negrin. In: Theory, Culture & Society, 17(5), 121-126.
Yu, LiAnne / Tng, Tai Hou (2003): Culture and design for mobile phones for China. In Katz, James E. (ed.): Machines That Become Us. New Brunswick, NJ: Transaction, 201-217.

Appendix

Table 1: Key accounts of mobile phone decoration

Decoration	Accounts	Total # of decoration explained	FGI #	Gender	# of decoration explained
Charms/straps	Favorite characters/ personal taste	11	1	Female	3
			3	Female	3
			4	Female	5
	Souvenirs/gifts	10	1	Female	1
			2	Male	1
				Female	4
			3	Female	1
			4	Male	1
				Female	2
	Matching with friends or boy /girlfriend	8	1	Female	1
			2	Female	3
			3	Female	1
			4	Female	3
	Convenience (e.g. easy to find)	3	2	Male	1
			3	Female	2
Stickers	*Purikura*	9+	1	Female	2
			2	Male	1
				Female	3+
			3	Male	1
				Female	1
			4	Female	1
	Favorite characters	4	1	Female	1
			2	Female	2
			4	Female	1

Prospective uses and mobile media

Communicative mobility and mobile work: the management of everyday life and communication networks in a mediatized world

Matthias Berg

Introduction

Concerning the interrelations of work, mobility and communication, research is in progress focusing on two very important fields: a group which could be called the 'mobile elite', exaggeratedly represented by the 'jet-set business society', and communicative relations in minority groups such as migrant and diasporic communities. At the same time, it seems that some decisive phenomena have been more or less excluded from scientific analysis so far: although sociology in particular is dealing with social processes and consequences of mobile labor, an expertise on related media use and aspects of communicative processes is widely missing in this discourse. That is why this article focuses on the *communication practices in situations of job-related mobility*, including individuals who have relocated due to job or educational reasons, are working in highly mobile business branches, or can be described as weekly commuters. My aim is to investigate to what extent different forms of local mobility correspond with specific patterns of communication-practices. As I will show, the appropriation of digital media is interrelated to a certain degree with the way people realize spatial requirements of an individualized labor market. Especially in terms of coordinating everyday life and maintaining remote relations, detailed communicative strategies are applied.

The core of this study applies the concept of *communicative mobility* (Hepp 2006, 2007, 2008). In general, this approach marks a tool to investigate the manifold interrelations between mobile people and media from a communication studies point of view by emphasizing two major aspects: firstly, the appropriation of increasingly mobile communication devices, and secondly, the growing orientation of stationary media terminals and services towards mobile users. Alongside, communicative mobility is intertwined with different forms of "local mobility", offering the opportunity to evaluate different forms of communication practices against the background of physical motion. As I will show, mobile devices are allowing spontaneous and flexible management of everyday life; however, stationary media do play an important role to support situative as well as biographic local mobility. Examples are the landline telephone and various internet-services, especially e-mail, messenger-programs, and social network sites. Additionally, the use of Skype for phone calls as well as instant messaging, StudiVZ or Xing[1] shows that not only portable devices but also mobility-oriented stationary services have to be considered when analyzing communication practices that support social relations in periods of increased work-related mobility.

The following chapter will provide a rough theoretical framework stressing the basic interconnections of communication, labor and mobility. Against this background I will shortly describe the concept of communicative mobility as a research perspective and then give a rough outline of literature dealing with mobility of labor and connected media appropriation. Finally, I will present some preliminary results of a research project that relates distinct communication practices to different types of local work-mobility by applying the concept of communicative mobility.

Theoretical contextualization

Combining the topic of "interrelations of mobility and media use" with my special focus on different forms of labor mobility leads to a vast set of theories dealing with post, second or late modernity and reflections on (compressed) time and space, as well as thoughts on networks, globality, flows and liquidity. What I want to do in this chapter is to narrow down these important approaches to the most essential theories. Even though this might be an all too coarse-grained effort, I will address three central topics: mediatization, the concept of mobile societies, and individualization.

The concept of *mediatization* describes the increasing diffusion of different forms of mediated communication into everyday life. Hereby, two major aspects are to be mentioned. From a quantitative angle, mediatization describes the increase of media technologies and their appropriation on a temporal, spatial and social level. From a qualitative angle, it describes the shaping power of media in the relations between (media)technological, communicative and sociocultural change (Hepp / Krotz 2007). One effect of mediatization, for example, can be seen in the blurring boundaries of everyday structures. Formerly separated spheres – such as family, friends, leisure time, school and work – increasingly overlap with each other against the background of an all-connecting web of media communication (Krotz 2005: 450 et seq.). Thereby, Krotz characterizes mediatization as a meta-process that – like individualization, for example – lies across other social phenomena and affects multiple areas of everyday life.

1 StudiVZ and Xing are social network sites based in Germany. While the former can be described as Facebook-equivalent primarily attracting university students, the latter is a business-platform comparable to LinkedIn.

Concerning (labor) mobility and corresponding communicative practices, mediatization brings up two issues. On the one hand, we have to ask to what extent a change of communicative patterns and forms might be connected to mobile work. On the other hand, the 'everywhere and any time' of our mediatized world has to be contextualized with social and cultural transformations concerning the notions of the professional as well as the private.

Among others, Tully and Baier show an increase of mobility in the 'western world' since the era of industrialization. When they are talking of *"mobile societies"* they mean that "at no point in history [...] the mobility of modern man was as extensive as today" (Tully / Baier 2006: 15). In their classical definition of mobility, they differentiate between social mobility, informational mobility and spatial mobility, which is the movement of a subject through a spatio-geographic system over time (ibid.: 34).

Urry, one of the most influential social researchers dealing with the mobility-issue, also points out to the accelerated development of mobility, especially within the last decades (2002, 2007). In his "Mobilities Paradigm" he even subsumes "five interdependent 'mobilities'": corporeal travel of people, physical movement of objects as well as imaginative travel, virtual travel and finally communicative travel (2007: 47). Furthermore:

> "Social research typically focuses upon one of these separate mobilities and its underlying infrastructures and then generalizes from its particular characteristics. This new paradigm by contrast emphasizes the complex assemblage between these different mobilities that may make and contingently maintain social connections across varied and multiple distances" (ibid.: 48).

What we can take from that is that in mobile societies, along with objects, goods, money, ideas etc., more and more people are in motion more often, with an increasing radius and for different reasons – for example, as migrants, tourists, when relocating or simply due to everyday movement like commuting. Against the background of mediatization they are given the opportunity to stay connected communicatively when in motion. Thus, the field of job-related forms of mobility and corresponding communication practices at least integrates two moments of Urry's request.

But what are the reasons for this increasing rate of mobility? One possible answer can be found in the process of *individualization* – especially since we are dealing with job-related mobility. For Beck and Beck-Gernsheim, individualization is manifested in the fact that "the *social significance* of inequalities has changed" (2002: 30). That means in short, "that in the wealthy industrial heartlands of the West" (ibid.) due to the structures of the welfare state, the destiny of an individual is no longer primarily determined by class, descent or status, but increasingly by his personal responsibility. Since education, job and career are more or less dependent on individual decisions, Beck and Beck-Gernsheim call the labor market the "'motor' of Individualization" (ibid.: 31).

Thus, individualization can be seen as the freedom as well as the burden of deciding for one's self, including "an *individualization of social risks*" (ibid.: 39). But being able to choose different career options goes along with being flexible in terms of location:

> "The labor market, by way of occupational mobility, place of residence or employment, type of employment, as well as the changes in social location it initiates, reveals itself as a driving force behind the individualization of people's lives" (ibid.: 32 et seq.).

Of course, the complexity of those three approaches is greatly reduced here. However, at this point their ability to explain and analyze for the concrete field of labor mobility and communication-practices is already sufficient: fostered by an individualized labor market, job-related mobility is constantly growing in its importance. This leads to an increasing number of formal (professional) as well as informal (private) contacts established during an individual's education and career – in other words, a pluralization of network-connections. Through mediatization, at least in its current dynamics constituted with the digital era, we now gain the opportunity of potentially keeping those networks connected.

As we can see, individualization, mobile societies and mediatization are braiding into a single logic strand. In order to operationalize those theories empirically, I will now introduce the concept of communicative mobility as an appropriate research-perspective for communication studies.

The concept of communicative mobility

The concept of communicative mobility, as developed by Andreas Hepp (2006, 2007, 2008), tries to offer an alternative research perspective concerning the appropriation of mobile media. Very roughly speaking, it aims to gain insight in the interrelations of mobility and media.

As the following illustration shows, communicative mobility can be split up into the branches of "'getting mobile' of communication devices" and the "'mobility focus' of stationary media" (cf. Hepp 2008: 137 et seq.). The first point can be compared to what is also labeled 'mobile media' and includes technologies like the mobile phone, notebooks, PDAs, mp3-players and other portable devices.[2] The mobility focus of stationary media refers to phenomena like camera surveillance of people on the move or media and services which grant the opportunity to stay in contact even across big distances (for example, via e-mail). Hepp also includes the traditional mass media here.

---

2 The examples Hepp gives here focus on technologies and their appropriation – a restriction that does not necessarily has to be adhered to. Media content also can be contextualized within the concept of communicative mobility (see Berg / Hepp 2007).

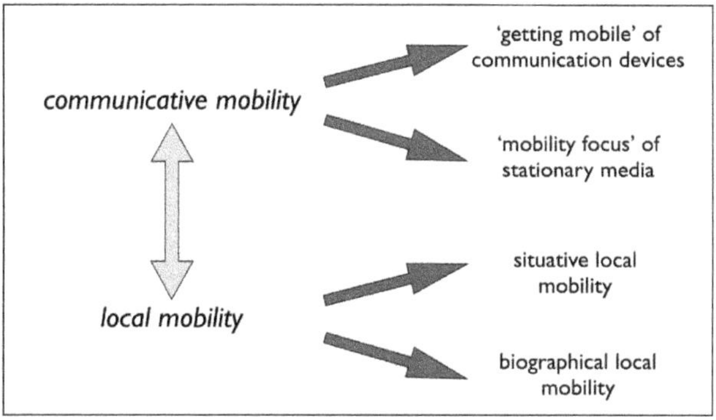

Communicative Mobility and Local Mobility (Hepp 2008: 138)

A further central feature of the model besides this conceptualization of communicative mobility is its juxtaposition with what Hepp calls "local mobility". Local mobility, like communicative mobility, is split into two categories: the so-called "situative local mobility", which describes the mobility of everyday life as in commuting to work, distances covered due to shopping, traveling or leisure activities, and "biographical local mobility", mobility during one's life-course like relocation or migration. In the sociology of mobility, those terms are paralleled by concepts such as "recurring" or "circular mobility", and "non-recurring spatial mobility" (cf. Schneider / Limmer 2008).

This model in general provides a structural differentiation – a grid, so to say – that allows us to conceptualize observations on forms of individual mobility, media use and their various possible interrelations. A central aim of the concept is to expand the understanding of mobile media or communication in academic research beyond simply focusing on portable devices such as the mobile phone. As Hepp states, this problem "is familiar from earlier media studies research [...] namely the separation of the technology itself from the broader social and cultural context in which it is used throughout the course of everyday life" (Hepp 2008: 131). Communicative mobility seeks overall to overcome this form of "technological centrism", specifically by contextualization as well as a critical research attitude (ibid.: 136 et seq.).

What we can take from this is that there is no full understanding of mobility and media without looking at the interrelations of mobile media, stationary media focusing on mobility and different forms of local mobility. Thus one central feature of communicative mobility is that it is inclusive while retaining a very practical character and direct applicability in diverse fields of research. Yet communicative mobility can not stand for itself: it has to be embedded within a wider theoretical framework matching the distinct field of research it is applied to. In terms of communicative mobility going along with different forms of job-related mobility we already have found mediatization, individualization and mobile societies to be useful.

Communication, mobility and labor – a brief outline of research-literature

At first sight, research literature dealing generally with mobility, labor and communication seems to be abundant. But a closer look reveals that only few approaches tend to integrate the varying aspects of mobility presented in the last two chapters. As it comes to different disciplines, sociology predominantly deals with the description and analysis of work-related mobility in the context of transformations of the labor-market, while the communicative momentum is missing or marginal. Otherwise, communication studies mostly investigate either mobile technologies in general or the consequences (digital) media have on professional practices as well as job-conceptions.

Communication, mobility and labor in sociological research

Recently, Schneider and Limmer have published their work on labor market and mobility (2008), contrasting different forms of 'mobile workers' with non-mobile employees. Hereby, their main emphasis is set on the motivations for and consequences of mobility in the individual and family-context. The researchers make clear, that there are two sides to mobility: "With respect to jobs, the decision to be mobile or not is primarily affected by two criteria: the threat of unemployment and the unique opportunities for advancement" (ibid.: 124). Additionally, as different as the motives are the strategies applied by individuals and families in negotiating the circumstances of mobility: beneath some positive effects – like job-improvement and sometimes a better financial situation – the negative side-effects are being stressed, such as "a significant negative impact on social relationships" (ibid.: 132) or a general disorientation (ibid.: 137). Resulting from their findings is a critical assessment of an over-simplified attitude towards mobility – be it in politics, economy or in a general societal discourse promoting "the altered social value of mobility" (ibid.: 119).

Bonnet et al. (2008) take the same line as Schneider and Limmer, also dealing with the compatibility and contrariness of (local) private and (mobile) professional lives. They, too, put very clear stress on the ambiguousness of labor mobility, stating that "There is no longer a clear-cut separation between choice and constraint; individuals sometimes do not know any longer if they do what they want or if they do what they are forced to" (ibid.: 159).

Deeply immersed into social analysis, what those two publications do not articulate are the communicative aspects of work-related mobility: if people are working away from their family and social environment, must not those aspects be regarded to be of central importance for managing absence? A slight exception to this rather common trend are the findings of Pelizäus-Hoffmeister (2001), who examines "social networks under circumstances of spatial mobility" and at least partly includes the communication habits of freelance journalists. She observes that as it comes to private social relationships, modifications of the criteria of interaction by means of changing forms of communication are a common reaction towards job-mobility (cf. ibid.: 139). But the willingness to substitute face-to-face interaction to a certain degree by mediated interaction essentially depends on the age of individuals: the younger the persons involved, so argues Pelizäus-Hoffmeister, "the bigger the willingness to use media for private purposes" (ibid.: 139 et seq.).

One of the most integrative approaches to modernity, mobility and communication is presented in a sociological examination of "mobility pioneers" (Kesselring 2006; Kesselring / Vogl 2008). The two central hypotheses of Kesselring and Vogl (2008) are "that the strong connection between geographical and social mobility is getting weaker" (ibid.: 163) and "that there is a conceptual change from the dominance of physical to virtual movement" (ibid.: 167). The authors underpin their empirical findings by three examples, each standing for a distinct type of "mobility management" (ibid.: 169):

- *centered mobility management:* a freelance journalist with strong personal networks – high physical mobility (daily commuting one hour to his office) and selective communication activities
- *de-centered mobility management:* a former broadcast manager, now working as an independent journalist, living on the Balearic Islands with a logistic home-base in Germany and an area for spiritual retreat in a Buddhist enclave in Southern Europe – high degree of international physical mobility as well as communicative connectivity
- *virtual mobility management*: a female top-level IT-journalist working half-day from her home, the rest of the day being a mother and housewife – no physical but high "virtual mobility" (ibid.: 169 et seqq.)

Central to this approach is the notion of mobility "*as an actor's competence to realize specific projects and plans while 'on the move'*" (ibid.: 167). This means that Kesselring and Vogl are not isolating different dimensions of mobility but that they are conceptualizing them as integral, with the effect that communicative, physical and social aspects are included in their typology of mobility management.
But what are the exact characteristics of "virtual movements" in the case of physical remoteness? What exactly is happening communicatively "while on the

move"? We will find some answers when moving on to media and communication studies.

Communication, mobility and labor in communication studies

How everyday, mediated interactions are shaped in partnerships determined by the influence of increasing professional and private mobility is shown by Dietmar (2008). Taking a communication-studies approach, she finds the ritual to be a key concept to describe communicative relations in partnership. Examples range from everyday coordination, spontaneous emotional support to good-night or good-morning-messages (119 et seqq.). Especially the latter two can be understood as a continuous re-confirmation of the relationship, as well as examples of the "phatic function of new technologies" (Tomlinson 2008: 67). Thus, Dietmar gives us an idea of how communicative coping with remoteness can work in partnership.

Next to this aspect of the private sphere, another focus of research which emerged in communication studies deals with the telework-issue. For example, Dimitrova (2005) aims to analyze distinct uses of media communication by employees in situations of remote work. She describes how the differences in experience and status of sales personnel in the telecommunications business create different attitudes and uses of media communication in the job. For instance, she finds ordinary sales representatives to be rather cautious towards the technology they have to use every day when doing business at the customer's site. Similar to O'Hara et al. (cf. 2003: 68 et seq.), Dimitrova stresses the importance of informal face-to-face relations with colleagues, and finds that their absence in telework can lead to discontent. However, she describes how sales executives have a rather positive attitude towards telework and connected media use, since they rely less on it but benefit from the opportunity of getting out of the stress- and interaction-zone in the office-environment.

Taking this analysis to a more abstract level, Dimitrova and O'Hara et al. focus on practical aspects of the mediatization of daily activities at work. In light of the rising influence of communication media on work life, it becomes clear that the very concepts of practical work as well as distinct self-conceptions are subjected to modification. A comprehensive change in the professional world is also the subject of the work of Wittel, who claims that "working practices become increasingly networking practices" (2008: 160). Even though the empiric evidence he gives is based on copresent ethnographic fieldwork in the British "New Economy", Wittel states that the "network sociality" he proclaims is "deeply embedded in communication technology, transport technology, and technologies to manage relationships" – it "is delocalized, it is a sociality on the move, a sociality over distance" (ibid.: 177).

As we can see, many fields of research are associated with the phenomena of mobility, work and communication. However, whereas sociological research deals with transformations of the labor market and the resulting consequences on an individual and societal level, communication research focuses on the connected communicative change; for example, at the workplace or in private life. Additionally, both disciplines are pervaded by a private-professional dichotomy, mostly concentrating on one aspect only.

Taken together, all the above mentioned results deliver a base one can start from when investigating the way job-related mobilities are managed communicatively. However, at the same time we have to conclude that at this point there are no substantial contributions dealing with this issue. Questions which remain unanswered are:

What role do mobile and stationary media play in the lives of 'mobile workers'? What does their communicative coordination of everyday life look like? How does the management of communication networks work?

In the next chapter I will try to provide some answers, presenting primary results of an ongoing research project on communicative mobility and work-mobility.

Communicative mobility and different forms of labor mobility – some first results

In this chapter I want to analyze to what degree different forms of job-related local mobility are articulated, in the way mobile as well as mobility-oriented media are appropriated. This means that some typical interrelations of local and communicative mobility will be discussed, while other communication practices – in this case the use of social network sites – will turn out to be less influenced by physical mobility.

The empirical data I am drawing on is taken from an exploratory research project,[3] addressing the overall question of how people whose everyday lives are influenced by job-related mobility appropriate digital media in order to maintain and extend their communicative relations. For that purpose, interviews were conducted with people from two urban areas in Germany (Frankfurt, Bremen and their surroundings respectively). The interviewees have jobs in different branches, and feature distinct mobility-strategies.[4] A thematic guideline was applied for the interviews including topics like career-, media- and mobility-

3 It is the qualitative part of my PhD-project I am currently carrying out at the Institute for Media, Communication, and Information (IMKI) in Bremen.

4 A total of 20 individuals between the ages of 20-35, with professions raging from industrial workers, teacher and police trainees as well as ordinary businesspeople, bankers and corporate consultants, were interviewed.

biographies, as well as everyday work routines, media use and communication networks. After word-by-word transcription, the interviews – together with scans of pre-structured communication-diaries filled out by the participants for one week – were analyzed following the guidelines of grounded theory.[5]

In this article I will refer to a sub-sample of three individuals who stand exemplarily for the following types of work-related local mobility:

- people who recently relocated due to changes in employment or education (no more than three months ago)
- long-term-commuters, returning home on weekends only (or even less often)[6]
- people working in mobility-intense businesses, not sleeping at home at least half the week.

Interrelations between local and communicative mobility

Analyzing differences and parallels concerning communicative mobility in terms of the above mentioned specifications of local mobility, two major aspects will be of dominant importance: the active communicative coordination of everyday life, and the maintenance of communication networks while "being away".

Communicative mobility and relocation

Moving is one of the most evident and "absolute" forms of (job-related) mobility. To relocate one's habitation, due to a new job or in order to attend university, for example, can be referred to as biographical local mobility. After one moves a certain distance away, this means that new social relations will be created at the new site that can be formal or institutional (e.g. colleagues, supervisors) as well as informal (e.g. friends and other acquaintances). However, maintaining social relations to the old home plays an important role. Furthermore, we have to consider that repeated relocations throughout a person's life course add up to her or his biography of mobility, leading to a distinct 'landscape of contacts' with social, communicative densifications. This becomes apparent with Tim, the first case of job-related mobility. Tim was born in a town on the Baltic Sea coast 28 years ago. This is where he went to school before he attended military service and then started studying informatics. After changing his course of studies, he moved to a university-town in eastern Germany. When interviewed, Tim had just moved to Bremen, where he is now

5 Additionally, the participants were asked to draw an open map of their individual communicative relations. Those drawings functioned as stimuli during the discussions.

6 Daily commuters have not been included since they do not necessarily have to use digital media in order to 'stay connected'.

working as a trainee teacher.[7] He is also an ambitious musician with numerous contacts spreading over the northern and eastern parts of Germany.

Tim is a good example for observing the attempt to communicatively connect to the former stations of a person's life after moving away: he is, more or less frequently, in contact with his family and former schoolmates, with some musicians at his first university, and with people from the city he just moved away from. Additionally, Hamburg and Berlin are urban centers which have attracted people who once belonged to the sphere of everyday face-to-face interaction. The media dominantly used to maintain relationships are typically numerous, and in Tim's case, include landline telephone (especially to contact the parents and also for "*people you haven't talked to for quite some time*"[8]), instant messaging (Skype's messaging-function mostly used "*by the way*"), and social software (StudiVZ). Another typical medium to stay connected is e-mail, but Tim does not really relate to that. In his case e-mailing shows the aspect of an integrating moment in form of a semi-official mailing list, which contains all fellow-trainees as well as his supervisor.

Overall, it becomes apparent that stationary media with a mobility-focus are employed to keep up communicative relations when relocating. The mobile phone as a portable device, on the other hand, is scarcely used to do so. Exceptions mostly refer to very close social ties – in Tim's case, the exchange of SMS with his girlfriend living in another city.

However, the central aspect of *portable devices*, in particular the mobile phone, refers to everyday coordination tasks. As it comes to relocating, the mobile phone turns out to be of high relevance during house-hunting and the period of transition from one residence to another, especially when a reliable stationary media infrastructure is not yet given. As Tim reports, his mobile phone bill went up during the last month (when moving) "*because right now I'm not at a fixed place and so I don't have a landline*". Besides those particularities, predominantly short-term adjustments and the prompt scheduling of meetings are negotiated via mobile phone calls and SMS. This especially includes meetings with friends and the communicative management of situative local mobility, such as commuting of short distances, traveling and visiting somebody. Tim's days, for example, are continuously alternating between his apartment, the school he teaches at and trainee seminars. Out of 20 mobile phone calls and SMS entered in his communication-diary, 15 refer to coordination of (mostly) recreational meetings or arrangements of visits. The importance of those management-aspects is reflected in Tim's perception of his

7 One has to notice here that this interview was recorded two weeks after moving, when the act of relocating still has a very situative character. The intention here is to collect communicative relations at a later point in time in order to eventually detect changes.

8 Citations from the interview-material are italicized.

personal media-ranking: he is absolutely convinced that "*without a mobile, I think life today would not work the way it actually does*".

Communicative mobility and weekly commuting

The decision for weekly commuting can be interpreted as a way of trying to negotiate diverging personal and professional spatial requirements. Consequently, weekly commuters leave their partner, family and / or friends behind during the week and return home on the weekends, some even less often. This means that they have two fixed centers of life which are connected by repetitive physical movement, in a form of situative local mobility. Accordingly, the central issue in terms of communication is to stay connected to those two poles of split corporeal presence, and can be expressed in a twofold challenge: on communicative maintenance in addition to everyday coordination at two alternating sites.

Our second example here is 33-year-old Marc, who lives and works in Frankfurt. He spent his life in the rural area where he was born, grew up, was a bank apprentice, and went to university, until he moved to the city in 2000 in order to work for a bank. Marc's job is rather well-regulated, with his own office and workplace fixed. He works in sales, offering financial products to corporate clients. However, he still drives home almost every weekend, where he stays at his parents' home and sees his friends. Within the last eight years Marc has not gotten to the point where he feels really connected to his urban environment.

Marc's media-setup is broad: he uses his private mobile phone, a landline telephone, and various internet-based services. E-mail is the most important channel to him, since he uses it at work as well as in private life. The telephone is also important; however, he differentiates between the landline, which is essential to his job, and the mobile phone, which he uses for private communication mostly on the weekends.

Compared to relocating, weekly commuting includes the same communicative features in terms of contact-maintenance: stationary but mobility-oriented media play a central role here. At the same time, we have to consider that the circumstances vary substantially in the point that weekly commuting means going back and forth between two fixed locations, whereas biographical local mobility produces places one does not regularly return to. As a result, one central difference is observable: for commuters, weekend activities are subject to intensive preliminary and mediated planning throughout the week. Suggestions are made and negotiated among friends or with the partner, often via e-mail or social network sites. But as the weekend approaches, the exact terms and arrangements are made using the mobile phone.

Communicative mobility and highly mobile work

Similar to commuting on a weekly basis, mobility-intensive professions represent a form of situative local mobility. The difference here is that employees are not alternating between defined spots, but visit changing locations, starting from and returning to their 'home base'. Work is project-oriented, including variable sites and durations. Consequently, the logic of stationary media for maintaining relations and mobile media for coordination tasks assigned to relocation and commuting dissolves. This extreme form of situative local mobility is mainly supported by devices getting mobile while stationary media are used on weekends only – if at all. Throughout the week, the (mostly corporate) mobile phone is functioning as the central means of professional as well as private interface, losing its emphasis concerning the arrangement of leisure time meetings. The reason is that there are usually no existing private face-to-face relations at the customer's site, and consequently no requirements for arranging such meetings. Instead, short-term arrangements give way to long-term planning of social activities.

In addition, highly 'mobile workers' report of immense time-consuming work-involvement, leaving almost no time for recreation. Thus, the few private moments left are used intensively for mediated interaction with one's center of life, in which social relations occur primarily face-to-face. Whenever possible, friendship networks are contacted, and so mediated communication serves to approve remote social relationships, especially if there is a partner at home. In addition to the mobile phone, e-mailing is the second dominant way of staying connected here. The experience of our third and final case, Lisa, offers an explanation for this:

Lisa is a 26-year-old corporate consultant working for a company with an office in Frankfurt. Together with her partner, she lives in a city nearby. She spends Mondays through Thursdays at customers' sites all over Germany, returning to the office each Friday. Her workdays mostly exceed twelve hours, which means that she is away and does not have much time for anything but work during the week. Thus, her major aims are to focus copresent social life on the weekends, and putting much effort into mediated interaction throughout the week whenever possible. The way she tries to 'stay connected' to her partner clearly implies a momentum of "ritual communication" described by Dietmar (2008): as her diary indicates, a short conversation with her boyfriend via mobile phone mostly marks the first as well as the last communicative activity on each weekday. But in addition, a high degree of situative work-mobility typically goes along with actively using times of transport (taxi, train, airport) either for working or spontaneous communication.[9] Also, Lisa reports that she

9 Compare Hulme and Truch (2005) for the influence of the mobile phone on what they call "interspace".

usually writes her boyfriend three to four short e-mails a day, even when in the office: "*I think that's a little bit normality, in order to keep up a normal relationship.*"

As she states, her smart phone and notebook with location-independent web-access via UMTS are clearly the most important technologies to her: "*Without laptop and blackberry really nothing works I think because […] you get e-mails all the time and you have to react all the time.*" The combination of mobile phone – in this case, enhanced with e-mail functionality – and UMTS-enabled notebook has become usual in mobility-intensive jobs. The aim is total communicative connectivity through 'infrastructure-autonomy'. In most cases the devices are provided by the employer who also takes care of the costs. Interestingly, Lisa has to pay a share of 20 percent of her phone bill anyway; in return, she is permitted to use the corporate device for private purposes. Therefore, she predominantly uses the blackberry for personal and business phone calls on weekdays, but unlike many colleagues, she switches her blackberry off on the weekends and replaces it with a private mobile phone. This kind of short-term and flexible local mobility is decisive for the predominant use of portable devices. Simultaneously, however, it clearly represents a very strong form of the "*'pressure' of communicative mobility*" described by Hepp (2008: 124 et seqq.).

Bringing together those observations, we can now say that different forms of local mobility actually are interrelated with different communication practices. I have compared three cases of 'mobile workers', stating that people in highly mobile jobs and commuters rather incorporate aspects of situative local mobility while relocating can be contextualized as a form of biographical local mobility. Commuting and relocated people are shaping their communicative environment by using stationary terminals predominantly for maintaining relations and mobile devices for the coordination of everyday life – for example, to arrange face-to-face meetings. In contrast to that, mobility-intensive work entails the use of mobile devices for all aspects of being away: communication in the job, as well as maintaining face-to-face and remote relations.

Of course, these results have to be understood as tendencies rather than absolute ascriptions. Exceptions do exist and can be explained by logic interferences: for instance, a recent relocation does not exclude situative mobility, and every situatively mobile individual has a distinct background of biographical mobility. Beneath individual habits and predispositions, different forms of mobility are just one – albeit important – factor influencing the appropriation of digital media. In the following chapter, we will see that the quality of appropriating social network sites, for example, is not directly influenced by types of job-related local mobility.

Stationary media in mobile situations – the example of social network sites

The emergence of social network sites can be seen as the latest substantial development in the context of 'new' digital media. Here I want to show that this phenomenon, too, can be contextualized within communicative mobility; however, the way those sites are appropriated – at least primarily – does not depend on individual forms of mobility.

As shown above, the three exemplary individuals are using similar media, but in different ways and for varying purposes (landline-telephones, at least one mobile phone, web-enabled computers for surfing, e-mailing and some sort of instant messenger). Simply speaking, all use mobile as well as (semi)stationary technologies and devices in order to stay connected. And even though none claimed social network sites to be indispensable, all use at least one: while Lisa and Marc are registered in the business-platform Xing, Tim as a future teacher decided not to do so, since he does not see how it benefits his career. He prefers the German Facebook-equivalent StudiVZ. Marc is also a member of Wer-Kennt-Wen ("who knows whom"), a network site popular in the mid-western parts of Germany. Additionally, he spends a respectable amount of time with online poker and role-playing games on the computer.

In terms of frequency, Lisa reports that she uses Xing once a week on average, mostly when she has to accept a contact-request. She has approximately 120 people in her contact list, among which are friends and personal acquaintances as well as business-contacts: "*Sometimes even clients add you, or so, and then you have the opportunity to contact them again, maybe.*" Still, Lisa mainly uses the site for checking on addresses and birthdates, like a "*personal address-book*".

Marc uses Xing even less often, about once a month. Half of his 40 contacts are friends and half are business-related. Contrary to Xing, the estimated 400 people he knows via 'Wer-Kennt-Wen' consist of contacts from his private sphere. Most of his friends are registered there and additionally he keeps on adding, and being added by, people he meets at parties and disco events. He uses the platform to arrange weekend-activities, as his communication diary indicates, as well as to 'rediscover' formerly known persons, former classmates or colleagues – a central aspect of sites like Wer-Kennt-Wen, he states.

Of the three interviewees, Tim most frequently uses a network-portal for interpersonal communication. Along with Skype and sometimes an additional instant messenger, StudiVZ belongs to the online-services he uses every day. He reports that he mainly exchanges e-mail-like personal messages with friends, and sometimes writes on pinboards, which are comparable to guest books on homepages. They can be read by every person who has access to the user's profile. Tim characterizes the pinboard-messages as "*rather extraneous*" or superficial, but at the same time appreciates receiving them, since it shows that "*the other person thought of me*". In short, Tim sees pinboards as a nice way to stay loosely connected. Out of the 120 persons in his contact list, he "*really directly deal[s] with maybe fifty or so. The rest simply is there because of social network somehow. And you just know each other if you have met, or something like that.*"

Now, what can we extract from this juxtaposition of different ways to appropriate social network sites? In part, it reflects the results of research published in this field so far. As Boyd and Ellison describe: "While networking is possible on these sites, it is not the primary practice on many of them, nor is it what differentiates them from other forms of computer-mediated communication" (2007: 2). Indeed, none of the interviewees ever mentions having made new acquaintanceships on a social network site:[10] "What makes social network sites unique is not that they allow individuals to meet strangers, but rather that they enable users to articulate and make visible their social networks" (ibid.). Hereby, the visualization can also include former social relations, as we have seen with the rediscovering- or reactivating-function of these sites. Additionally, a function of social network sites is their use as an address-book (cf. Renz 2006), which brings to mind Wittel's reflections on the role of databases in the network-sociality-context (2008: 167 et seq. and 178). Furthermore, social network sites are also to be perceived in the very sense of communication platforms with different forms of communication converging (e-mail-similar functionality, guestbook-like pinboards, special messengers, etc.).

Another central topic present in the appropriation of social network sites refers to the private-professional dichotomy: despite their official portrayals as either business or leisure platforms, the way social network sites are being used is not as clearly formal or informal at all. One dimension of this issue has already been denoted in the examples: business-platforms typically include personal as well as professional contacts (cf. Renz 2006). Concerning the latter ones, this allows the institutionalization of professional relations on an individual level. It is also of interest where and when 'business-platforms' are accessed – some interviewees do it at home, others at work. At the same time, many report, that they check their private portals, e-mails and SMS when in the office, Marc, for example:

> *"Well I'm honest there, since this is anonymous, every now and then I log in to an online role-playing game there [at the office] you know? [...] because if you need five minutes every now and then for relaxing and then, then you simply check on something private in the internet."*

10 This clearly contrasts Marc's experience concerning his poker- and role-playing-communities where he got to know people and even exchanges private issues – but still stressing that he would not call them friends.

Concerning the interrelations of mobile work and communication social network sites can help to cope with what could be named a 'mutual spatial dispersion' of individuals. It allows institutionalizing business-contacts on a personal level as well as (potentially) staying connected to single people and networks that have been generated throughout a single person's history of mobility. Contact-lists are usually fed by all sorts of sources resulting from biographical as well as situative local mobility: former school-mates, friends from university and internship-colleagues, as well as current co-workers and even clients. Thus, the quality of single relationships seems to be highly variable even in the longest contact-lists, ranging from very close friends – where online platforms are an additional way to stay in contact next to other media – to more loose connections and in some cases innumerous collections of non-recurring meetings. As a result, social network sites can be understood in terms of having a mobility-focus as stationary media. At the same time, there is no empirical evidence so far that supports clear interrelations between different types of job-related mobility and corresponding appropriations of social network sites.

Conclusion and perspectives for future research

In this article I argued that processes of individualization are a key factor in accounting for work-related mobility in general. The latter can be seen to be associated with the pluralization of communicative networks in mediatized mobile societies. In this context, I applied the concept of communicative mobility as a specific research perspective for communication studies in order to examine how different types of job-related local mobility are managed through specific appropriation of media technologies.

First, relocating, weekly commuting and highly mobile forms of occupation were characterized as predominantly either situative or biographical local mobilities. Furthermore, the correlating challenges those situations implied for the individuals were stressed – maintaining relationships to former environments, integrating two centers of life and staying connected to one place while constantly being on the move.

Second, distinct ways of dealing with those mobility-forms communicatively were assigned. Indeed, specific media preferences and practices emerged, concerning the coordination of everyday appointments and maintaining remote relationships. These two main functions are realized by certain strategies of using communication devices "getting mobile" as well as stationary terminals with a "mobility focus". But it is important to understand those two dimensions as conceptualizations of communicative mobility rather than a fixed classification of technologies, since a simple description of online-communication like voice-over-IP, messengers and especially e-mailing as being stationary is becoming problematic while UMTS-enabled notebooks, wireless local area networks and blackberries are gaining importance.

Third and finally, we heard about the importance of stationary media by means of the recent developments of social network sites. Those are nurtured by contacts resulting from situative as well as biographical local mobility. The manifold opportunities network sites offer range from address book-functionality and the filing of acquaintances (as a modern form of the business card) to converging communication tools adding up to existing media. However, concerning the way those sites are actually appropriated on an individual level, direct references linking back to work-related mobilities are not evident yet. Thus, more research is required, since strategies of appropriation are at the moment massively in the process of forming and are by no means stable.

Additionally, the findings on social network sites denote another issue that must be considered when doing research on the entanglement of labor, mobility and communication: namely, the private-professional dichotomy, which still determines various academic perspectives. Today's research in this field tends to concentrate on the consequences of labor-mobility and communication in terms of either private or professional spheres. However, such a binary construction is dissolving verifiably, which Krotz has shown to be a consequence of mediatization (cf. 2005: 450 et seq.). As a result, investigations have to grow more integrative.

References

Beck, Ulrich / Beck-Gernsheim, Elisabeth (2002): Beyond status and class? In: Beck, Ulrich / Beck-Gernsheim, Elisabeth: Individualization: Institutionalized Individualism and Its Social and Political Consequences. London: Sage, 30-41.

Berg, Matthias / Hepp, Andreas (2007): Musik im Zeitalter der Digitalisierung und kommunikativen Mobilität: Chancen, Risiken und Formen des Podcastings in der Musikindustrie. In: Medien & Kommunikationswissenschaft. Sonderheft 1: Musik und Medien, 28-44.

Bonnet, Estelle / Collet, Beate / Maurines, Béatrice (2008): Working Away from Home: Juggling Private and Professional Lives. In: Canzler, Wert / Kaufmann, Vincent / Kesselring, Sven (eds.): Tracing Mobilities. Towards a Cosmopolitan Perspective. Aldershot: Ashgate, 141-162.

Boyd, Danah. M. / Ellison, Nicole B. (2007): Social network sites: Definition, history, and scholarship. In: Journal of Computer-Mediated Communication 13/1. URL: http://jcmc.indiana.edu/vol13/issue1/boyd.ellison.html [17.02.2008].

Dietmar, Christine (2008): „Wir telefonieren jeden Abend ... das ist uns ganz wichtig." Rituale bei der mediatisierten Kommunikation in Paarbeziehungen. In: Thomas, Tanja (ed.): Medienkultur und soziales Handeln. Wiesbaden: VS, 105-118.

Dimitrova, Dimitrina (2005): Work Relationships and Communication in Distant Work. In: Nyíri, Kristóf (ed.): A Sense of Place. The Global and the Local in Mobile Communication. Vienna: Passagen Verlag, 283-291.

Hepp, Andreas (2006): Kommunikative Mobilität als Forschungsperspektive: Anmerkungen zur Aneignung mobiler Medien- und Kommunikationstechnologie. In: Ästhetik und Kommunikation 37/135, 15-22.

Hepp, Andreas (2007): Kommunikative Mobilität in der Diaspora. Eine Fallstudie zur kommunikativen Vernetzung der türkischen Minderheiten-Gemeinschaft. In: merz 6/2007, 35-46.

Hepp, Andreas (2008): Communicative Mobility after the Mobile Phone: The Appropriation of Media Technology in Diasporic Communities. In: Hartmann, Maren / Rössler, Patrick / Höflich, Joachim R. (eds.): After the Mobile Phone? Social changes and the development of mobile communication. Berlin: Frank & Timme, 131-152.

Hepp, Andreas / Krotz, Friedrich (2007): What 'Effect' do Media have? Mediatization and Processes of Social-Cultural Change. Paper presented at the ICA Conference 2007, San Francisco, USA.

Hulme, Michael / Truch, Anna (2005): The role of interspace in sustaining identity. In: Glotz, Peter / Bertschi, Stefan / Locke, Chris (eds.): Thumb Culture. The Meaning of Mobile Phones for Society. Bielefeld: Transcript Verlag, 137-147.

Kesselring, Sven (2006): Topographien mobiler Möglichkeitsräume. Zur sozio-materiellen Netzwerkanalyse von Mobilitätspionieren. In: Hollstein, Betina / Straus, Florian (eds.): Qualitative Netzwerkanalyse. Konzepte, Methoden, Anwendungen. Wiesbaden: VS Verlag, 333-357.

Kesselring, Sven / Vogl, Gerlinde (2008): Networks, Scapes and Flows – Mobility Pioneers between First and Second Modernity. In: Canzler, Wert / Kaufmann, Vincent / Kesselring, Sven (eds.): Tracing Mobilities. Towards a Cosmopolitan Perspective. Aldershot: Ashgate, 163-180.

Krotz, Friedrich (2005): Mobile Communication, the Internet, and the Net of Social Relations. In: Nyíri, Kristóf (ed.): A Sense of Place. The Global and the Local in Mobile Communication. Vienna: Passagen Verlag, 447-457.

O'Hara, Kenton / Brown, Barry / Perry, Mark (2003): Mobile Work, Technology and Place. In: Nyíri, Kristóf (ed.): Mobile Democracy. Essays on Society, Self and Politics. Vienna: Passagen Verlag, 65-72.

Pelizäus-Hoffmeister, Helga (2001): Mobilität: Chance oder Risiko?: Soziale Netzwerke unter den Bedingungen räumlicher Mobilität. Opladen: Leske und Budrich.

Renz, Florian (2006): Networking im Internet-Zeitalter: Neue Praktiken des online-gestützten Netzwerkens und ihre gesellschaftlichen Auswirkungen. In: soz:mag 09. URL: http://sozmag.soziologie.ch/09/sozmag_09_renz.pdf [03.02.2009].

Schneider, Norbert F. / Limmer, Ruth (2008): Job Mobility and Living Arrangements. In: Canzler, Wert / Kaufmann, Vincent / Kesselring, Sven (eds.): Tracing Mobilities. Towards a Cosmopolitan Perspective. Aldershot: Ashgate, 119-140.

Tomlinson, John (2008): "Your Life–To Go": The Cultural Impact of New Media Technologies. In: Hepp, Andreas / Krotz, Friedrich / Moores, Shaun / Winter, Carsten (eds.): Connectivity, Networks and Flows: Conceptualizing Contemporary Communications. Cresskill: Hampton Press, 59-68.

Tully, Claus J. / Baier, Dirk (2006): Mobiler Alltag. Mobilität zwischen Option und Zwang – Vom Zusammenspiel biographischer Motive und sozialer Vorgaben. Wiesbaden: VS.

Urry, John (2002): Mobility and Connections. URL: www.ville-en-mouvement.com/telechargement/040602/mobility.pdf [02.02.2009].

Urry, John (2007): Mobilities. Cambidge & Malden: Polity Press.

Wittel, Andreas (2008): Towads a Network Sociality. In: Hepp, Andreas / Krotz, Friedrich / Moores, Shaun / Winter, Carsten (eds.): Connectivity, Networks and Flows: Conceptualizing Contemporary Communications. Cresskill: Hampton Press, 157-182.

The iPhone and mobile access to the internet

Rich Ling & Pål Roe Sundsøy

Introduction

The marketing of the iPhone has been a type of watershed event in the world of mobile communication. The sleek device has captured the attention of the public and the design and functionality of the iPhone has been seen as something new in the rather more established world of mobile phones. It is being marketed using unique forms of promotion; its design is notably different than that of other mobile phones and, of most interest to this analysis, it is being used to access the internet more regularly than other types of mobile phones. In addition, as we show here, it is encouraging the use of the mobile internet.

The iPhone, particularly in its first release, represented several new dimensions in mobile communication. The phone was only sold via one operator (AT&T) and there was the explicit notion that the device would be "locked" to that net. While there have been other mobile phones models that were released only by a particular operator, this is perhaps the only example of a whole brand that was reserved in this way.

The marketing of the iPhone and the ability of Apple to make a media event out of the commercialization was capitalized on in several ways. Apple was able to enter a contract with AT&T whereby Apple would receive a sum for each handset sold plus a certain amount per month for the duration of the customer contract. The idea was that the "buzz" surrounding the device would result in new customers – with their accompanying 2-year contracts – for AT&T. Indeed, it was reported that more than half of the iPhone users were new subscribers to the AT&T net.[1] It was reported that about 270000 phones were sold during the first days of its release meaning that AT&T had received as many 135000 new subscribers (Benderoff 2007).

Beyond the marketing aspects of the iPhone, the device itself offers some new wrinkles in the use of a mobile phone. The functions of the phone included the simple design, the "multi-touch" and relatively high-resolution screen that provides sharp images and an "accelerometer" that automatically reconfigures the screen from landscape to portrait format when the phone is rotated. On the negative side the first version of the iPhone was relatively slow when compared

1 http://news.cnet.com/8301-10784_3-9747031-7.html

to other 3G phones and was not necessarily seen as an advanced business phone but rather as more as a multimedia device (Macedonia 2007).

Mobile communication is often seen as a technology of the intimate sphere (Ling 2008). When, for example looking at the number of calls, text messages and other "events"[2] generated by a mobile phone, the interpersonal activities dominate. Indeed an analysis of the traffic on the network for this operator during the spring of 2008 shows that 90 % of the events are calls or text messages. Internet traffic accounts for only about 5 % of the events. Further, analysis shows that half of the calls we make and text messages we send go to only a handful of other persons (Ling 2008). Thus, the mobile phone is, and has been a communication device that we largely, though not exclusively, use among to keep in contact with our nearest friends and family.

The iPhone may be changing this. It has widely been reported that Google recorded 50 times as many searches via the iPhone than with any other handset.[3] This is a remarkable statistic that, if true, may well indicate that some users are establishing a praxis of using the iPhone for activities beyond the more routine calls and text messages. Indeed, with a certain amount of hyperbole, commentators such as Jonathan Zittrain suggest that the iPhone and other such devices are fundamentally changing the way that the internet functions (Zittrain 2008).

The Google statistic is interesting, but it is only one application. There are, of course many other types of internet functionality that are not captured in that type of an analysis. While we are not able to discuss the specific applications being used, we are interested in the broader question of absolute use of the iPhone using the telephone network (not the WiFi network) to access the internet. The fact that a relatively large number of iPhones were informally imported onto the net meant that their use patterns could be traced in the traffic data.[4]

This introduces the research question that we wish to examine, namely just how much do iPhone users access the internet? There is material that is available that will allow us to answer this question. The broader issue, however, is why this should be the case. Three dimensions can be drawn on when considering this latter issue. These are the usability of the iPhone, the degree to which there is "background" contact with the net and finally whether using the net via the iPhone is a type of self-fulfilling prophesy.

The iPhone does indeed have a unique interface that distinguishes itself from other mobile phones. As noted, it has a large touch screen and there is an emphasis on simple interaction. In addition, the iPhone facilitates web access through simple, intuitive web access. Another possible explanation is that the

2 An event is when the user logs onto the net.

3 http://www.macworld.co.uk/ipod-itunes/news/index.cfm?rss&newsid=20446

4 Individuals could, for example, purchase an iPhone and log onto the alternative network where it could be "cracked" and used on any GSM network.

iPhone runs "background" applications that appear as web activity regardless of the intentions of the owner. In this respect, the owner may have configured the device to perform a function without realizing the consequences. In an odd way, this has assigned agency to the device (Bakardjieva 2005). Another twist on this is that increased web browsing may be the result of a self-fulfilling prophesy (Merton 1968). In this case, the buzz associated with the iPhone as an advanced device somehow eggs on the nascent owner to increase his / her surfing. The analysis here will only be able to determine whether the iPhone encourages web use. Based on that result it is possible to speculate as to the reason for use.

Method

The data comes from the database of a large Norwegian telephone operator before the operator officially sold the iPhone. The material indicates that as of the second quarter of 2008 there had been about 8000 iPhones used on the net. About 1500 of these had been "cracked" and then, in all likelihood, sold to another owner. These were not included in the analysis. Of the remaining users, approximately 1000 iPhone users were included in the analysis.

The analysis shows that there is an over-representation of young male iPhone users. The phone used previous to the iPhone is quite often a high-end Nokia phone (E65 or N95) but there are also a surprising number of basic phone users who move to the iPhone. The analysis also indicates that iPhone users consume more "IP" traffic than do others.

The data for this analysis came from anonymous records derived from actual traffic data. A total of 3917 users were included in the sample, the 1000 iPhone users and about 2900 users of other phones. They are divided into four categories based on their terminal (handset) type as of the second quarter of 2008.[5] We also compared the iPhone users to three other user groups.

The final sample included 957 Apple iPhone users, 989 Nokia E65 users, 985 Nokia N95 users and 986 general users. Data on the sample's use of voice, SMS, MMS, WAP and internet in the second quarter of 2007 – that is immediately previous to the release of the iPhone – and during the second

5 The material was gathered using the operator's traffic database. The names and identities of the users are anonymous and there is no possibility of tracing the material back to a particular person. The analysis was conducted under the broader license to examine the use patterns in order to better develop the network. In this respect, knowledge of internet use represents a new use pattern that will have implications for the development of the network. The material on terminal type and traffic statistics are recorded in the general traffic database.

quarter of 2008 was gathered.[6] In addition, this approach allowed us to examine the data on the "previous" handset type. Clearly, the iPhone users had another type of handset in the second quarter of 2007 since the iPhone had not been released. We did a superficial examination of the Nokia N95 and E65 users and the broad majority of them had the same type of handset in T1 and T2.

The specific internet based information was the number of times that the individual logged onto the internet from their phone. This was compared to see the change in behavior between 2007 immediately before the introduction of the iPhone and a year later in 2008. The mean change in the number of "events" that is the number of times per month that an individual logged onto the net is 32.75 for the whole sample.[7] In addition, we were able to compare the number of bytes downloaded in September 2008. The data for the previous year was not available so only a cross-sectional analysis was possible in this case.

This data extraction allowed for both a longitudinal and a cross sectional analysis comparison of use. The Nokia phones were selected since at that point they were recognized as being advanced full function devices that would be in competition with the iPhone. In addition, the general sample of users provided insight into how the use of the iPhone (as well as the Nokia devices) was different from the use of a broad group of users who used a random sample of other terminal types.

As noted above, the iPhone was not sold outside the US during the time covered by this analysis. The persons who have an iPhone during this period purchased it in the US. This involved downloading new firmware into the phone and circumventing the security protocols of Apple.[8] Once imported, the device was "cracked" to allow its use on a local network. To unlock an iPhone a local SIM card was inserted and thus the device was registered on the network of the operator. This points to both the popularity of the phone and to the fact that the device was not commonly available at the time of the analysis. This also means that the individuals who adopted the iPhone cannot be seen as normal users. Rather they are in all likelihood those who can travel to the US to purchase a phone and who have the technical savvy to crack the phone, or at least know people who can assist them.

---

6 The data reported here is the mean monthly traffic for the particular type of traffic. That is the quarterly data divided by three.

7 It needs to be noted that the sample examined here is very special and should not be generalized to the whole population.

8 The next generation of the iPhone it has become available outside the US and can be legitimately bought so as to avoid the need for the dodgier process of thwarting the security software of the phone.

Findings

General analysis of iPhone users

The material gleaned from the traffic data indicates that while there are a limited number of iPhones on the network[9], they were used in a significantly different way. In this section, we will examine the influx of iPhones to the network and the socio-demographics of the users. We will examine the types of subscriptions that the users employed and the types of handsets that they used before owning an iPhone. Finally, we will examine the main question regarding their use of the iPhone for surfing the internet.

There is material on the age and gender of the person having the subscription in the traffic database. This material is generally correct, but in about 10 % of the cases, the wrong age is given and the same can be said of the gender. This comes from cases where, for example a 40-year-old father subscribes for his 14-year-old daughter. In that case, the "age" of the subscription owner is 40 but the behavior is that of a 14-year-old. As noted, this occurs in about 10 % of the cases. Thus, there is an inherent level of error in the data.

Looking at the material for the iPhone we can see that there are significantly more males who own one. When comparing the iPhone to general users, it is clear that the iPhone is a device that is fancied by male users.[10] While the general population of users is roughly 52 % male vs. 48 % female, the split was 84 % vs. 16 % for the iPhone. When compared to two other advanced phones, the Nokia E65 (a business phone)[11] and the N95 (a "high-end" phone) significantly more males had the E65 and significantly fewer had the N95[12] when compared to the iPhone.

Looking at the age distribution of the different phones, it is clear that the iPhone appeals to younger users. Looking across these handsets, the mean age of the Nokia E65 users was 47.8, the Nokia N95 users was 46.5 and the mean age of the general users was 48.6 years. The mean age for the iPhone users was 35.6. This difference is significant.[13]

An analysis of the type of subscription indicates that a large number of the iPhone users – almost 60 % – had a subscription associated with their job. The data shows that 50 % of the E65 and 64 % of the N95 users had a business subscription. By contrast, only 29 % of the general users had a business subscription. This is an important finding since the users of these subscriptions often do not pay for their own use. This would free them from directly feeling

9 Far less than one percent of all handsets.
10 Chi2 (3) = 364.40, sig < 0.001
11 Chi2 (1) = 12.14, sig. < 0.001
12 Chi2 (1) = 10.491, sig. < 0.001
13 f (3,2880) = 149.48, sig. < 0.001

the economic consequences of surfing the internet, a practice that can often add significantly to the cost of using the device.

Two thirds of the persons who made the transition to the iPhone in this period were using a Nokia phone before adopting the Apple product. The material indicates that the owners of Nokia handsets were less loyal than the owners of other types. Nokia owners were somewhat more likely than others to make the transition to the iPhone. Owners of Sony Ericsson and Samsung handsets are a smaller portion of the total mobile phone population. They were, however somewhat less fickle in their loyalty.

When looking at specific handset types, the largest number of persons transferring to the iPhone came specifically from the Nokia E65 and Nokia N95. Previous to having an iPhone 16 % of the iPhone users had an E65 and 15 % had an N95. These are both advanced handsets that offer a variety of functions. In this way, the multi-functionality of the iPhone would match well with many of the functions found on these two devices. As we have already seen there are a significant number of males who own these Nokia models, though the age comparison is different for the Nokias and the Apple phone.

By contrast, it is surprising to look at the second tier of devices that were traded in for an iPhone. These were three older Nokia phones that were actually no longer in production. These phones, the Nokia 3510 (5 % of all phones replaced by the iPhone) 3410 (2 %) and 6310 (2 %) are phones with a black and white screen no camera and only limited ability to access the internet. That this group would make the transition to the use of the iPhone is startling. While it is difficult to understand the issue, it is possible to speculate that the users were well entrenched in the menu structure of the older phone and perhaps reluctant to change to another device that would be difficult to learn. In this way, the simplicity of the iPhone allowed them to switch phones with the least trouble.

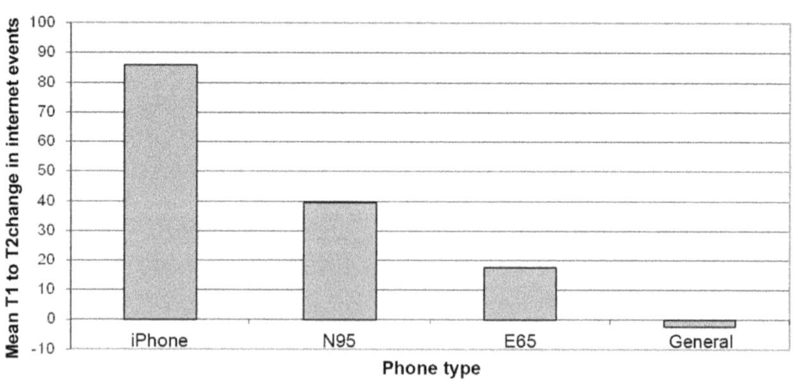

Figure 1: Mean change in internet monthly number of internet "events" between T1 and T2

Traffic analysis

When looking at the actual change in surfing behavior the analysis shows that there was an increase in the mean number of internet "events" recorded for the iPhone users. The iPhone users increased the number of times that they logged onto the net by a mean of just under 90 times a month. That is, they showed an increase of about three times a day compared to the period before they had the iPhone. For the users of the Nokia N95 the increase was about 40 times a month, the E65 users increased the number of times they logged on by just under 20 and the general users actually reduced the number of times that they logged on in the period between T1 and T2.[14] As the material in figure 1 suggests, the users of the N95 also increased the number of events.[15] Users of the N95 recorded an increase in the number of internet events significantly more than either the users of the E65 or the general users. Finally, the analysis shows that the users of the E65 logged more events than the general users.[16]

Thus, while it is true that the iPhone was clearly that device that most encouraged surfing the internet, it is not the only one. Users of both the Nokia N95 and the E65 also showed an increase in the number of times that they logged onto the internet using their phones when compared to general users.

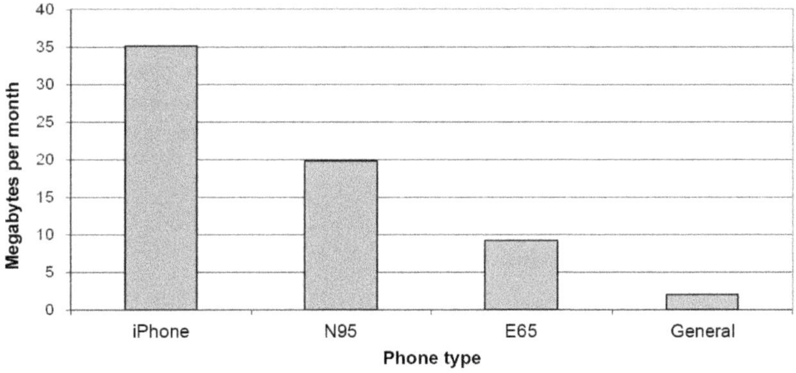

Figure 2: Megabytes down loaded per month by different types of phones, September 2008

14 When all four handset conditions were included, the results were f (3, 4169) = 35.885, sig < 0.001. In one on one comparisons, increase in the number of events for the iPhone was significantly more than for the N95, the E65 and the general population at less than the 0.001 level of significance.

15 As with the iPhone, the users of the N95 increased the number of events by a significant level when compared to the users of the E65 and the general users. In both cases this was at less than the 0.01 level of significance or better.

16 This is at less than the 0.001 level of significance.

The different handsets were also compared in a cross-sectional analysis of the number of bytes that were downloaded over a month's time. This analysis is shown in Figure 2. Again, users of the iPhone clearly downloaded more data than did users of the other phones.[17] The users of the iPhone downloaded approximately 35 megabytes or about one megabyte per day during the month. Users of the N95 downloaded approximately 20 megabytes per month, the Nokia E65 users downloaded just under 10 per month and the general users downloaded a mean of about 2 megabytes per month or about 0.06 megabytes per day. This last statistic is perhaps a surprisingly high number when considering all the general users of mobile telephony. However, the data shows that of all the general users, over half never used any internet traffic.

Looking at the data a slightly different way, 13 % of the iPhone users never used their phones for data traffic while just under half of the general users never did. The data shows that 18-20 % of the of the two Nokia phones users never used their phones for downloading data traffic. These numbers were almost reversed when looking at those who downloaded more than 100 megabytes per month. Just about half of the iPhone users downloaded more than 100 megabytes where only 5 % of the general phone users did the same. The material also shows that 28.6 % of the N95 users and 18.3 % of the E65 users downloaded this much data.

When compared to the E65 and to the general users there was a clear difference between the use of iPhone and these other types of users. When compared to the users of the N95, that is in many respects also a multi-media device, the results were still statistically significant, but the margin was not as pronounced.[18] With all the other two-way comparisons, the results showed a high degree of statistical significance.[19]

Discussion and conclusion

In summary, the users of the iPhone use the internet more than other users. This is not simply a cross-sectional finding, but the analysis shows that the same users altered their behavior because of the adoption of the iPhone. Why would this be so? What is it about the iPhone that would encourage users to change their pattern of use? There are several explanations that are available. These include the socio-demographics of the iPhone users, the nature of the iPhone when compared to other devices and finally the nature of the subscriptions that the users had.

17 Looking across all four handsets, the ANOVA showed that the results were $f(3,3996) = 22.166$, sig. < 0.001.

18 $F(1, 1998) = 6.586$, sig $= 0.01$

19 In all the other comparisons, the level was near to or below the 0.001 level.

The analysis shows that there is a somewhat different socio-demographic group that has adopted the iPhone, namely younger males. It is interesting to look at this by comparing it to material describing the use of the PC based web. There is not a particularly high percentage of young males who report surfing the net on a daily basis when compared to other age groups (39.4 % for 18 to 30 year olds vs. 40.5 % for 31 to 49 year olds) By contrast, young adult females do not report surfing as often as middle-aged adult females (37.2 % for 18 to 30 year olds vs. 51.8 % for 31 to 49 year olds). However, the picture is somewhat different when looking at the time used by those people who do surf the web (PC based). In that case, as with the demographic profile of iPhone users in this study, young adult males who report surfing the PC based web also report using the web for longer time periods than other groups. Looking at males 18 to 30 year old, they reported using the web a mean of 118 minutes per day vs. 77 minutes for 31 to 49 year old males. The 18 to 30 year old females reported using the web a mean of 88 minutes per day vs. 50 minutes for 31 to 49 year old females. Thus, it is possible to suggest the correspondence between iPhone users, who were more over-represented among young adult males, and the PC based surfing behavior of the same group.

At the same time, the data suggests that the same users surfed more after they had started using the iPhone when compared with their own previous use. This points to the idea that there was something about the actual surfing experience on the iPhone that encouraged the surfing. Here it is important to look at the differences between the iPhone and other mobile phone handsets on the market.

The screen of the iPhone is quite large, there is an enhanced touch screen that allows for multiple touch points (used to enlarge or reduce sections of the screen image) and the software allows the user to "flick" through images. Thus, there is a different experience associated with using the iPhone when compared to the more static experience of other phones. These qualities may provide the user with a better experience of the internet than other devices that have smaller screens and less advanced displays.

It is also possible to suggest that the accessibility of the internet entices users to link up to the internet more often than with other phones. That is, there may be fewer steps in the process of accessing the internet. With the elimination of each step, the likelihood of accessing the net is greater.

Another possible explanation is that access to net-based e-mail and other background services might mean that whether or not the user actively uses the device to link to the internet, there is the generation of internet traffic. That is, the phone is configured to check into different web based applications as a type of background task. This functionality is not unique to the iPhone. Other devices can be configured in this way. However, the fact that the iPhone strives to give users access to a variety of web-based services, this type of configuration is likely a common one.[20]

Putting this into a social perspective, by configuring the device to use these background services, the users is assigning a type of agency to the device (Bakardjieva 2006). In an interesting way, this is making the machine into a social actor, or at least an economic actor. The device has been configured in order to make decisions for the individual. In spite of the fact that the individual does not have a specific overview over the individual actions of the device, they nonetheless have economic consequences. To the degree that this takes place, there will be the generation of web traffic.

The increase use of the iPhone on the net can also be the result of a self-fulfilling policy (Merton 1968). Apple and the various telecom operators who have sold the iPhone have focused on the iPhone as a mobile web terminal (Becker 2007). To the degree that owners accept this characterization, they will be encouraged to surf from the iPhone in order to justify their purchase.

Finally, the data indicates that as many as 40 % of all iPhone user had a job that paid for their use. This means that there was not the same economic constraint as experienced as "private" users. Again, this would perhaps encourage the use of web-based services.

All told, several factors contribute to the increased use of the iPhone for surfing the web. However, the T1 / T2 setup of this analysis indicates that there was a particularly strong effect of the device itself. Users of the N95 were active users of the mobile internet and this indicates that there is a general drift in this direction, particularly among that avant-garde mobile users. Nonetheless, the iPhone has shown itself to be a device that encourages use of the internet in a way that was not common until now.

References

Bakardjieva, Maria (2005): Internet society: The internet in everyday life. London: Sage.

Bakardjieva, Maria (2006): The consumption junction revisited: Networks and contexts. In: Kraut, Robert / Brynin, Malcom / Keisler, Sara (eds.): Computers, Phones, and the Internet. Oxford: Oxford University Press, 97-108.

Becker, David (2007): iPhone Web Traffic Beats Windows Mobile. URL: http://blog.wired.com/gadgets/2007/12/iphone-web-traf.html [16.12.2008].

20 The iPhone considered in this analysis was the first version. The "apps" store had not become available and so there were a more limited number of web-based services available.

Benderoff, Eric (2007): Apple credits iPhone buyers: Early adopters of the device who are upset over quick price cut get $100 compensation. Chicago: Tribune.

Ling, Rich (2008): New Tech, New Ties: How mobile communication is reshaping social cohesion. Cambridge: MIT Press.

Macedonia, Michael (2007): iPhones Target the Tech Elite. In: COMPUTER, 40(6), 94-95.

Merton, Robert (1968): Social theory and social structure. New York: Free Press.

Zittrain, Jonathan (2008): The Future of the Internet. And How to Stop It. New Haven: Yale.

About the authors

Matthias Berg studied cultural studies, musicology and economics with an emphasis on popular culture as well as media and communication research. He is a member of the Institute of Media, Communication, and Information (IMKI) at the University of Bremen, Germany. Currently Matthias Berg is working on his PhD-project which focuses on the interrelations of (labour)mobility, mediated digital communication, interpersonal relations, and social communication networks.

Leopoldina Fortunati studied literature, philosophy and social science. She then worked at the University of Padua, Venice, Trieste and Udine. She is professor of Sociology of Communication with a focus on new media at the University of Udine. Her research interests include gender studies, cultural processes such as fashion, and the role and the social use of communication and information technologies, in particular fixed and mobile phone as well as internet.

Richard Harper is Principal Researcher at Microsoft Research in Cambridge and co-manages the Socio-Digital Systems group. Richard is concerned with how to design for 'being human' in an age when man-as-machine type metaphors, deriving from Turing and others, tend to dominate thinking in the area. Trained as a sociologist and with a strong passion for ordinary language philosophy, he has published over 100 papers and is about to publish his 10th book, Texture: Human expression in the age of communication overload, (MIT Press). Amongst his prior books is the IEEE award winning The Myth of the Paperless Office (MIT Press, 2002), co-authored with Abi Sellen. His work is not only theoretical or sociological, but also includes the design of real and functioning systems, for work and for home settings, for mobile devices and for social networking sites. Numerous patents have derived from his work. Prior to joining MSR, Richard helped lead various technology innovation and knowledge transfer companies, while in 2000 he was appointed the UK's first Professor of Socio-Digital Systems, at the University of Surrey, England. It was here he also set up the Digital World Research Centre. Prior to this he was a researcher at Xerox PARC's fifth lab, EuroPARC, in Cambridge. He completed his PhD at Manchester in 1989. He lives in Cambridge with his wife and three troublesome but occasionally delightful children.

Joachim R. Höflich studied economics, social science and communication sciences. He then worked at the universities of Augsburg, Hohenheim, Bamberg and Munich. Since September 2002 he is Professor for Communication Science with a focus on media integration at the University of Erfurt, Germany. Research interests include media use and media effects; ‚new' communication technologies and changes of mediation cultures; the theory of (technically) mediated (interpersonal) communication; intercultural comparative media communication and mobile communication. Latest publications: Mobile Communication in Everyday Life: Ethnographic Views, Observations and Reflections (edited together with M. Hartmann). Berlin 2006; After the Mobile Phone? (edited together with M. Hartmann and P. Rössler), Berlin 2008.

Georg Florian Kircher is a PhD Student at the University of Erfurt, Germany. He studied communication sciences and political sciences and worked at the University of Erfurt. His recent research focus is on media adoption and usage in everyday life, mobile media and media convergence, concepts of mobile lifestyle and Habitus, and ethnographical fieldwork.

Amparo Lasén, Ph.D. studied sociology at the university Paris V-La Sorbonne. She is associate professor of Sociology at the University Complutense de Madrid, Spain, and a member of the research group Cibersomosaguas of the same university (www.cibersomosaguas.com). Her research focuses on the social implications of mobile media, especially in relationship with affectivity and the configuration of contemporary subjectivities, from everyday practices and uses to political collective actions. Prior to her current position, she was the Vodafone Surrey Scholar at the DWRC of Surrey University, where she conducted cross-cultural research on mobile phone uses and practices; and had been academic visitor at the Department of Sociology of the LSE.

Rich Ling, PhD, University of Colorado in 1984, is a professor at the IT University of Copenhagen and at Telenor's research institute located near Oslo, Norway. He has also been the Pohs visiting professor of communication studies at the University of Michigan in Ann Arbor, Michigan and now he holds an adjunct position in that department. He is the author of the book: New Tech, New Ties: How mobile communication is reshaping social cohesion (MIT). He is also the author of a book on the social consequences of mobile telephony entitled The Mobile Connection: The cell phone's impact on society (Morgan Kaufmann). In cooperation with Jonathan Donner he has written the book Mobile Phones and Mobile Communication. Along with Scott Campbell he is the editor of The Mobile Communication Research Series and he is an associate editor for The Information Society.

Christine Linke is a communication and media researcher, focusing on the impact of social relationships. In her dissertation thesis at the University of Erfurt, Germany, she has been analysing the process of mediatization in the everyday life of couples. Her research areas of interests include communication in social relationships, everyday life communication practices, media and the construction of identity, and ritual media communication.

Isabel Schlote studied communication sciences and literature and now is a PhD candidate at the University of Erfurt, Germany. Her research emphasis is on interaction strategies, media usage and contextual behaviour in everyday life. In her PhD project she is looking at the communication structures and situational aspects in public transport. She also focuses on the use, possible enhancements and combinations of qualitative research methods.

Satomi Sugiyama, Ph.D., is assistant professor of Communication and Media Studies at Franklin College Switzerland. She has been conducting research on the way young people perceive and use mobile communication technologies in various cultural contexts. Her work has appeared in several collections of essays and a journal, including Sugiyama, S. (2009) Decorated mobile phone and emotional attachment for Japanese youths (Electronic Emotion, the Mediation of Emotion via Information and Communication Technologies, edited by J. Vincent & L. Fortunati, Peter Lang) and Katz, J. E., & Sugiyama, S. (2006). Mobile phones as fashion statements: Evidence from student surveys in the US and Japan. New Media and Society, 8 (2), 367-383. She completed her Ph.D. in Communication at Rutgers University in 2006, and has been a member of Center for Mobile Communication Studies at Rutgers University, International Communication Association, and National Communication Association.

Pål Roe Sundsøy (Msc Physics, NTNU, UNSW 2004) is a Data Mining Specialist at Telenor Group Business Development & Research located near Oslo, Norway. He has earlier worked in the Hinode Space mission project at the Institute of Theoretical Astrophysics at University of Oslo, Group of Applied Physics at University of Geneva, and World Health Organization. His current research interests include social network analysis, large-scale data processing and visualization and understanding of complex quantitative information in various mobile markets.

Jane Vincent graduated in Social Sciences from the University of Leicester and worked for over twenty years in the UK mobile communications industry designing, developing and implementing new products and services. She is Visiting Fellow with the Digital World Research Centre, for whom she has delivered numerous industry funded studies and publications on the social practices of mobile communications users since joining them in 2001. Jane is due to complete her Doctorate in 2010 with the University of Surrey on the topic of emotion and mobile phones and has recently published jointly with Leopoldina Fortunati an edited volume with Peter Lang Berlin, Electronic Emotion, the Mediation of Emotion via Information and Communication Technologies. Jane is UK representative on the COST Action 298 european project researching participation in broadband society.

www.ingramcontent.com/pod-product-compliance
Ingram Content Group UK Ltd.
Pitfield, Milton Keynes, MK11 3LW, UK
UKHW021832100926
8556IPUK00004B/292

9 783631 597651